Playing God with Emerging Technologies

Playing God with Emerging Technologies

How to Avoid the Traps of Techno-Optimism and Techno-Pessimism

Moti Mizrahi

BLOOMSBURY ACADEMIC
NEW YORK • LONDON • OXFORD • NEW DELHI • SYDNEY

BLOOMSBURY ACADEMIC
Bloomsbury Publishing Inc, 1359 Broadway, New York, NY 10018, USA
Bloomsbury Publishing Plc, 50 Bedford Square, London, WC1B 3DP, UK
Bloomsbury Publishing Ireland, 29 Earlsfort Terrace, Dublin 2, D02 AY28, Ireland

First published in the United States of America 2025

Cover design: Kathi Ha
Cover image © 3DSculptor/iStock

A catalog record for this book is available from the Library of Congress.

ISBN: HB: 978-1-6669-6452-3
ePDF: 979-8-2162-6321-0
eBook: 978-1-9787-6097-4

Typeset by Deanta Global Publishing Services, Chennai, India

For my boys
Eli and Jonathan

Contents

Preface

I have been teaching applied ethics, specifically Artificial Intelligence (AI) Ethics, Bioethics, and Engineering Ethics, at the Florida Institute of Technology since 2015. During this time, I have given a lot of thought to the role that "playing God" arguments can and should play in debates over the ethical design, development, deployment, and use of new or emerging technologies. Roughly, "playing God" arguments go something like this: "We should not play God, so we should not use this new or emerging technology because developing, deploying, or using it amounts to playing God." Such arguments are often used when the ethical use of new or emerging technologies is in question. For example, on solar geoengineering, Joanna Haigh, Professor of Atmospheric Physics at Imperial College London, says, "to play around with the climate system like that—playing God—I think is just foolish" (McSweeney 2019). Typically, "playing God" arguments are interpreted as warnings against interfering in God's creation or nature. But this interpretation of "playing God" arguments as a call to non-interference with God's creation or nature may be too narrow. When arguers make such arguments, they are concerned about interfering with God's creation or nature, to be sure, but they have additional moral concerns that the non-interference with nature interpretation of "playing God" arguments leaves out.

In this book, then, I argue that "playing God" arguments should be construed, not as warnings to leave God's creation or nature alone, but rather as invitations to think carefully about all the ways in which the design, development, deployment, and use of new or emerging technologies could go seriously morally wrong. According to my interpretation of "playing God" arguments, such arguments invite us to think carefully about all the ways in which the use of new or emerging technologies could go seriously morally wrong, given the fact that we are not as powerful, wise, and good as God is supposed to be. On my view, then, "playing God" arguments are more a critique of humanity than an appeal to anything supernatural like God. They are pleas to acknowledge human limitations, to avoid hubris and eventual humiliation, and to be cautious and humble instead. For this reason, on my interpretation of "playing God" arguments, such arguments cannot be dismissed simply by raising doubts about the existence of God.

Construed in this way, I argue, "playing God" arguments can have a legitimate role to play in debates concerning the morality of new or emerging technologies, contrary to those who argue that such arguments are redundant, unhelpful, or even meaningless. In fact, on my interpretation, "playing God" arguments could help to inform the governance of new or emerging technologies. In this book, I develop and outline a conceptual apparatus (i.e., an argumentation scheme with accompanying Critical Questions) for analyzing and evaluating "playing God" arguments as they are made in debates over the ethical development and deployment of new or emerging technologies, such as gene editing, in vitro fertilization, artificial wombs, social media, and the like. As such, this book may be valuable to scholars and students (both undergraduate and graduate students) of the social, moral, and legal aspects of science and technology, particularly those who work in ethics and applied ethics, as well as to general audiences concerned with moral and ethical problems arising from advancements in science and technology.

Part I of this book sets up the theoretical framework for my interpretation of "playing God" arguments. First, I explain what "playing God" arguments are, when they are typically made, and how they are usually interpreted in the context of debates over the morality of new or emerging technologies. Then, I put forth my interpretation of "playing God" arguments and the role that the concept of God should play in such arguments. Finally, I explain how to analyze and evaluate "playing God" arguments and give examples from debates over the ethical development and deployment of emerging technologies where such arguments are commonly made. This part of the book expands and elaborates on the argumentation scheme for "playing God" arguments put forward in my article, "How to Play the 'Playing God' Card," which was published in *Science and Engineering Ethics* in 2020.

Part II of this book consists of use cases where "playing God" arguments have been made or could be made. These use cases go beyond the familiar biomedical technologies, such as IVF and artificial wombs, about which "playing God" arguments are often made. These use cases include the following: climate engineering or geoengineering, Artificial Intelligence or AI, autonomous vehicles or self-driving cars, and Lethal Autonomous Weapon Systems or LAWS. Each use case is meant to be a standalone chapter that can be used as a case study in an applied ethics course, such as Technology Ethics, Engineering Ethics, AI Ethics, and the like.

In that respect, this part of the book drew much inspiration from the valuable experiences I had as a faculty advisor to Florida Tech's teams that competed in Lockheed Martin's Ethics in Engineering Case Competitions in 2023 and 2024. In this competition, teams of two undergraduate students, accompanied by a faculty advisor, present their solutions to hypothetical cases that pose problems with ethical, technological, business, and engineering dimensions. The conceptual apparatus developed in this book (namely, an argumentation scheme for "playing God" arguments with its accompanying Critical Questions) is meant to provide a general framework through which moral problems in technology ethics can be addressed. As such, this conceptual apparatus allows us to avoid the extremes of techno-optimism and techno-pessimism as well as develop a properly critical attitude with respect to new or emerging technologies.

Part I

"Playing God" Arguments in Theory

1

The Ethics of New or Emerging Technologies

"I feel bad sometimes that I ever did it." That is what John Sylvan told *The Atlantic* during an interview about his enormously successful invention, namely, the single-serve brewing pod, which is also known as the Keurig K-Cup (Hamblin 2015). When he came up with the K-Cup idea, Sylvan had no idea that it would catch on so quickly or that it would be widely used in many households across America. He thought that pod-based coffee machines would be used in offices only. Nowadays, however, many American homes have pod-based coffee machines. The problem is that K-Cups are causing enormous environmental damage because they are not recyclable or biodegradable. In fact, the US Securities and Exchange Commission (SEC) charged Keurig with making misleading statements about the recyclability of its single-use coffee pods, or K-Cups, and the company agreed to pay a fine of 1.5 million dollars (Prentice and Heavey 2024).

Could Sylvan have foreseen the catastrophic impact his invention would have on the environment? It is difficult to say. Perhaps he lacked sufficient knowledge at the time to suspect that his invention would turn out to be disastrous for the environment. *Should* Sylvan have at least considered what sort of impact his invention might have on the environment? This question seems somewhat easier to answer. Even if he *could* not have foreseen all the ways in which his invention would be disastrous to the environment, it seems reasonable to say that Sylvan *should* have thought about the ways in which his invention could be used or misused. Indeed, the fact that he now regrets ever inventing the single-serve brewing pod suggests that, by his own lights, he should have done better.

Even if we *can* do something due to advancements in science and technology, it does not necessarily mean that we *should* do it. As Dr. Ian Malcolm, the character played by Jeff Goldblum in the movie *Jurassic Park* (1993), famously put it, "Your scientists were so preoccupied with whether or not they could, they didn't stop to think if they should" (quoted in Spence 2008, 100). In other words, the scientifically and technologically feasible is not always the morally right thing to do. Even if advances in science and technology make something feasible, we need to think carefully about whether it is the morally right thing to do. For instance, thanks to recent advances in the science and technology of genetics, we may be able to bring back to life extinct species by means of genetic engineering and cloning (Appleton et al. 2024). But *should* we bring back to life extinct species? In the movie *Jurassic Park* (1993), it did not turn out so well.

Thinking carefully about the morality of new or emerging technologies, from design and development to deployment and usage, is the domain of technology ethics. Technology ethicists apply the sort of careful and systematic thinking philosophers are trained in, specifically, *logical argumentation*, to moral problems and ethical issues about technology in general and specific technologies in particular. Moral problems and ethical issues arise when advancements in science and technology enhance our power to act in ways we were unable to act before and allow us to make choices we were unable to make before. For example, before the advent of in vitro fertilization (IVF), people who suffered from infertility did not have the power to have their own children even if they wanted to. Nowadays, IVF gives infertile people the ability and the opportunity to have their own children, provided that they can afford it, since eggs and sperm can be used to create fertilized eggs (embryos) in laboratory dishes.

When advancements in science and technology grant us the power to do things we were unable to do before or make choices we were unable to make before, we naturally feel uneasy as legitimate moral and ethical concerns arise. Consider, for example, what was said about the radio in 1924 (Abbott 1924, 466–7):

> Here is a new *power* inconceivable in magnitude. How is it to be *controlled*? How operated? How supported? Is this another *monster* which humanity as a reckless Frankenstein has created without *controlling* it? (emphasis added)

Note that these questions express moral concerns about the power of the new invention and our ability to control it (i.e., "How is it to be *controlled*?") as well as the sophistication of the new invention, our ability to understand it, and our knowledge to operate it properly (i.e., "How operated? How supported?").

These questions express moral concerns that can arise with respect to many new or emerging technologies. That is, we can, and should, ask the following questions with respect to almost any new or emerging technology.

- Can we control it?
- Do we understand it?
- Will we be corrupted by it?

These are fundamental questions about the limits of our power to control our environment, our knowledge of our world and our place in it, and our corruptible nature or moral character.[1]

Similar moral concerns can be raised, and have been raised, about almost any kind of new or emerging technology. The psychologist, Amy Orben, calls this the "Sisyphean Cycle of Technology Panics" (Orben 2020). The first stage of this cycle, which Orben calls "panic creation," is the stage at which "a society [is] becoming worried about a new technology" (Orben 2020, 1146). A society's worries about a new or emerging technology can be characterized as moral worries or ethical concerns partly because the new or emerging technology itself is often characterized as being either *good* or *bad* (or both) for society. Like *right* and *wrong*, *just* and *unjust*, and *fair* and *unfair*, *good* and *bad* are moral categories, which are the sort of things that moral philosophers or ethicists, including technology ethicists, think about carefully and systematically.

To illustrate the point about how a moral panic is created within a society, Orben (2020) gives the example of the internet. According to Orben (2020, 1146):

> When the Internet increased in popularity, [. . .] many analyses [. . .] adopted stark *utopian* or *dystopian* points of view [. . .]. On the positive side, the technology was often construed as providing informative and educational content to children, leveling the social playing field, or providing better access

to information. Yet, there were also concerns about access to inappropriate content, promotion of violence, or child exploitation (emphasis added).

This kind of *utopian* versus *dystopian*, or *good* versus *bad*, dichotomy leads to and reinforces *optimistic* and *pessimistic* attitudes toward technology.

Those who tend to emphasize the overall value of technology in general and extol the benefits of specific technologies in particular can be characterized as *technological optimists*. Technological optimists often recognize and acknowledge the bad consequences of technological innovation, but they insist that humanity is better off with it and would be worse off without it overall. For techno-optimists, technological advancement has mostly improved our living conditions and further technological advancement will continue to do so. On the other hand, those who tend to question the overall value of technology in general and point to the risks of specific technologies in particular can be characterized as *technological pessimists*. Technological pessimists often recognize and acknowledge the good consequences of technological innovation, but they insist that humanity is not better off with it than without it overall. For techno-pessimists, technological advancement has had a negligible effect at best and a negative effect at worst on our living conditions.

A recent example of techno-optimism in action comes from the digital technology industry, more specifically, the Artificial Intelligence (AI) sector. Some top executives in the AI industry are self-professed techno-optimists. For example, the CEO of OpenAI, Sam Altman, is a self-professed techno-optimist. He believes that AI technologies, such as his company's chatbot, ChatGPT, will usher in a bright future for humanity and that we should do whatever we can to bring about this bright future as soon as possible. At the very least, we should not put any obstacles in the way of AI progress. In Altman's own words, "I think AI is good. [. . .] I think it's going to be awesome. [. . .] I think it's going to be great. [. . .] I think we're heading towards the best world ever" (Roose 2023).

Another key player in Silicon Valley, which is an epicenter of the digital tech industry in general and the AI industry in particular, is Marc Andreessen. Andreessen wrote a "Techno-Optimist Manifesto" with a focus on AI technologies. In his manifesto, Andreessen (2023) labels the claim "We are told to be pessimistic" about technology as a lie. I suppose he means to say that those who say that we should be pessimistic about technology are lying to us, not that it is a lie that we are being told to be pessimistic about technology, which would make the claim "We are told to be pessimistic" false. Accordingly, Andreessen (2023) asserts that techno-optimists believe that there is no problem "that cannot be solved with more technology." As for AI technologies, Andreessen (2023) claims that they "can save lives" and that "any deceleration of AI will cost lives." He goes on to identify "tech ethics" as "the enemy" of techno-optimists and accelerationists, that is, those who want to see the development and deployment of AI accelerated rather than decelerated.

Although he asserts without argument that AI *can* save lives, Andreessen is probably right about that. AI technologies are already deployed in the healthcare industry to assist in drug development, disease diagnosis, health monitoring, digital health consultation, managing medical data, and more. For example, a deep learning model called Deep Learning based Automatic Detection (DALD), which was developed by researchers from the Seoul University Hospital and College of Medicine, is used to analyze chest radiographs and detect malignant cell growth, such as potential cancers (Nam et al. 2018). Other potentially life-saving applications of AI technologies include detecting sepsis (Cech 2022) and more (Floridi 2023, 142–67).

Just as AI technologies *can* save lives, they *can* also harm, destroy, and even take lives. First, AI technologies can cause, and have already caused, harm to human lives. For example, social media platforms, such as Facebook and Instagram, whose recommendation algorithms are driven by AI, are said to be harmful to the mental health of users, especially the mental health of young users. One study has found that posting edited (by means of AI-powered filters, such as beauty filters and augmented reality) photos of oneself (or so-called "selfies") on social media is associated with body-image issues and disordered eating (Cohen, Newton-John, and Slater 2018). Second, AI technologies can destroy, and have already ruined, human lives in a variety of ways. For example, the mother of a Dutch sixteen-year-old in Amsterdam who was profiled by an AI-enabled risk profiling system called "ProKid 12 - SI" said that "It messed up his life" (Knowles 2024). This is because risk profiling systems, such as ProKid, "create new risks in practice" (La Fors 2020, 6), including privacy violations, false positives, biased results, and more.[2] Finally, AI technologies can take, and have already taken, human lives. For example, in the ongoing war between Russia and Ukraine, which started when Russia invaded Ukraine in 2022, both sides have already deployed AI-enabled drones that are capable of autonomous flight and target acquisition, thereby reducing the need for humans in the loop (Satariano and Mozur 2024). As one Ukrainian soldier told reporters, "We could sit in the trench drinking coffee and smoking cigarettes and shoot at the Russians" (Satariano and Mozur 2024).[3]

The fact that AI technologies, like almost any other technology, can be deployed to do good as well as cause harm shows that techno-optimism and techno-pessimism are not tenable positions with respect to the ethics of technology. That is, technology as such is neither wholly good nor wholly bad morally speaking. Some technologies can be used to do morally good things. Other technologies can be used to do morally bad things. Even the same technology can be used to help as well as cause harm to others. Techno-optimism seems to encourage an uncritical adoption of new or emerging technologies, whereas techno-pessimism seems to encourage an uncritical rejection of new or emerging technologies. Instead of techno-optimism and techno-pessimism, I submit, we need to assume a critical attitude with respect to new or emerging technologies. Martin Peterson (2020, 177) describes this critical attitude toward new or emerging technologies as an intermediate view between techno-pessimism, which is too cynical, and techno-optimism, which is too naive. After all, some technologies warrant more praise than criticism, for example, medical devices like pacemakers, whereas other technologies warrant more criticism than praise, for example, destructive devices designed to cause death and destruction on a massive scale like nuclear, chemical, biological, and radiological weapons (also known as weapons of mass destruction or WMDs). To adopt a critical attitude about specific technologies, we need logical arguments.

One type of argument that is frequently made when the ethical design, development, deployment, or use of a new or emerging technology is in question is known as "playing God" arguments. In the next chapter, I explain what "playing God" arguments are and how to interpret them. Then, in Chapter 3, I take an argumentation theory approach to "playing God" arguments. More specifically, I sketch an argumentation scheme for "playing God" arguments together with its accompanying Critical Questions, which build on the argumentation scheme first outlined in Mizrahi (2020). In the last chapter of Part I of the book, Chapter 4, I show how to analyze and evaluate "playing God" arguments using this argumentation scheme with its accompanying Critical Questions. Together, the argumentation scheme for "playing God" arguments and its accompanying Critical Questions are meant to provide a conceptual apparatus with which "playing God" arguments can be analyzed and evaluated. In that respect, this framework is meant to provide a constructive lens through which to determine whether we need to make sure that we are able to control, understand, and benevolently use a new or emerging technology before it is morally permissible for us to deploy and use it. As such, it allows us to avoid the extremes of techno-optimism and techno-pessimism and develop a properly critical attitude with respect to some new or emerging technology.

Notes

1 I return to this point in Chapter 2 where I discuss "playing God" arguments.

2 I say more about AI technologies in Chapter 6.

3 I say more about Lethal Autonomous Weapon Systems (LAWS) in Chapter 8.

2

"Playing God" Arguments

Should we bring extinct species back from extinction? The geneticist, Eriona Hysolli, says that we should. She works at Colossal Biosciences, which is a de-extinction company that attempts to bring back the woolly mammoth by means of gene editing and Assisted Reproductive Technologies (ART). On the other hand, the biologist, Elizabeth Hadly, and the legal scholar, Deborah Sivas, say that we should not. Hadly and Sivas make the following argument against de-extinction (Hysolli, Hadly, and Sivas 2023):

> We are woefully unprepared—legally, morally, and scientifically—to "*play God*" by resurrecting extinct species. Although the "gee whiz" factor (and the financial incentives) might push us in that direction, the better use of society's resources would be to work on saving what we still have (emphasis added).

As their use of the phrase "play God" suggests, Hadly and Sivas put forth a "playing God" argument against the use of new or emerging technologies, such as genetic engineering and ART, to bring back species that are long extinct.

"Playing God" arguments are fairly common in technology ethics.[1] As Ronald Sandler (2014, 15) notes, "Research on emerging technologies, particularly genetic technologies, is sometimes described as 'playing God.'" When the phrase "playing God" is used in debates over the morality of new or emerging technologies, such as gene editing or genetic engineering, it is usually interpreted as a warning against interfering with God's creation or nature. For example, John Evans (2002, 125) takes the phrase "playing God" to stand "in for the argument that HGE [i.e., Human Genetic Engineering] is an arrogant *interference with the nature that God has created*" (emphasis added). In other words, the phrase "playing God" is understood as a call to non-interference with God's creation or nature. As John Weckert (2016, 88) puts it, "playing God is *interfering with nature or modifying nature*, as in the genetic engineering example, and that this is morally wrong" (emphasis added). Weckert (2016, 97) goes on to say that "playing God" is "expanding the sphere over which we have control to the extent that we no longer have the knowledge or capacity to competently make decisions with respect to the expanded part." Likewise, according to Ted Peters (2007, 173), "By 'playing God', [. . .] [w]e mean changing nature."

For the most part, textbook treatments of "playing God" arguments also construe such arguments as warnings to refrain from interfering with God's creation or nature. For example, in the context of the debate over prenatal genetic testing, the "playing God" argument against trait selection is presented as follows in one textbook (Barry 2012, 239):

> It is frequently argued, religiously, that trait selection is against God's will for human reproduction and life's meaning and purpose. Because children, regardless of traits, are a gift from God, by *intervening in the divinely directed course of reproduction*, as by trait selection, *we are playing God* and, therefore, denying God's wisdom of what is best for us (emphasis added).

In other words, to attempt to select the traits of newborns is to intervene in human reproduction, which is best left for God or nature, or so the argument goes.[2] As the author of another bioethics textbook puts it, those who put forth "playing God" arguments "fear that humans are going beyond appropriate limits in remolding or 'recreating' the human's nature" (Veatch 2016, 146). Along similar lines, Hans-Jürgen Link (2013, 444) interprets "playing God" arguments as arguments that express "the worry [. . .] that [synthetic biologists] might 'play' with nature in a carefree way" (cf. van den Belt 2009).

This interpretation of "playing God" arguments as a call to non-interference with God's creation or nature may be too narrow. When arguers who are concerned with the morality of new or emerging technologies make such arguments, they may be concerned with interfering with God's creation or nature, to be sure, but they may also have additional moral concerns that the non-interference with nature interpretation of "playing God" arguments leaves out. So, an alternative interpretation of "playing God" arguments is needed. Taking an argumentation theory approach, I first explain the role that the concept of God should play in "playing God" arguments in this chapter. Then, in Chapter 3, I sketch an argumentation scheme with accompanying Critical Questions, which capture the moral concerns expressed by "playing God" arguments, along the lines first outlined in Mizrahi (2020). The accompanying Critical Questions are designed to help us determine whether developing, deploying, or using a new or emerging technology would count as "playing God" when a "playing God" argument is made. If my interpretation is plausible, then "playing God" arguments should be construed, not as warnings to leave God's creation or nature alone, but rather as invitations to think carefully about all the ways in which the use of new or emerging technologies could go seriously morally wrong.

On my interpretation, "playing God" arguments can have a legitimate place in debates concerning the morality of new or emerging technologies, such as gene editing, contrary to those who argue that such arguments are redundant, unhelpful, or even meaningless. For instance, Weckert (2016, 98) writes:

> [T]he "playing God" phrase is redundant and perhaps just a rhetorical device [. . .]. If all of the work is done by the *consequences* of an action and not by its intrusion into the realm of God or nature, then *playing God adds nothing* to cost-benefit analysis, the precautionary principle, or some other risk-assessment strategy. From a secular perspective then the notion of playing God is redundant (emphasis added).

For Weckert (2016, 98), then, "the 'playing God' phrase is redundant" because he thinks of "playing God" arguments as arguments from consequences. In my argumentation scheme for "playing God" arguments, however, the work is done by the concept of God, not by the consequences of actions.[3]

Along similar lines, Thomas Cavanaugh (2002, 119–20) argues that "reference to God is not necessary [and] not helpful in bioethical disputes." Others have argued that the phrase "playing God" is meaningless. For instance, according to Philip Ball (2010), "playing God" is a meaningless "cliché mouthed mindlessly." If my interpretation of "playing God" arguments is plausible, however, then the concept of God can be meaningful and useful in disputes over the morality of new or emerging technologies provided that "playing God" arguments are interpreted, not as warnings to leave God's creation or nature alone,[4] but rather as invitations to think carefully about all the ways in which the use of new or emerging technologies could go seriously morally wrong.

To appreciate how "playing God" arguments can be helpful and meaningful in debates over the morality of new or emerging technologies, we first need to understand the role that the concept of God should play in "playing God" arguments. So, I now turn to explaining the role that the concept of God should play in "playing God" arguments.[5] Clearly, the concept of God is supposed to play some role in "playing God" arguments. After all, why mention God at all if the concept of God plays no role whatsoever in such arguments? But what role *should* this concept play? How *should* we decide what counts as "playing God"? More specifically, what determines whether the design, development, deployment, or use of a new or emerging technology amounts to "playing God"?

The traditional conception of God is that of a maximally perfect being (Mizrahi 2013). More explicitly, "the orthodoxly conceived monotheistic god of traditional Western theism" is an "omniscient, omnipotent, omnibenevolent, eternal creator *ex nihilo* of the universe" (Oppy 2006, 16). That is, the concept of God is that of an all-powerful (omnipotent), all-knowing (omniscient), and all-good or morally perfect (omnibenevolent) being (Wierenga 1989, 1).[6] As an omnipotent being, God's power is maximal, which is why God is supposed to be the sort of being that would never lose control over anything, including new or emerging technologies. In other words, any new or emerging technology, no matter how powerful, is no match for God's power. For this reason, new or emerging technologies could not get out of control when God is in charge of them. Unlike God, human beings are not omnipotent, and so they could lose control over new or emerging technologies when they design, develop, deploy, or use them. For this reason, new or emerging technologies could get out of control when humans, not God, are in charge of them.[7]

Likewise, as an omniscient being, God's knowledge is maximal, which is why God is supposed to be the sort of being that would be able to understand any new or emerging technology and foresee all the consequences (good or bad) that might come about from developing, deploying, or using that technology. In other words, any new or emerging technology, no matter how sophisticated, is no match for God's wisdom. For this reason, new or emerging technologies could not be misunderstood or have unforeseen consequences when God is in charge of them. Unlike God, human beings are not omniscient, and so they may be unable to understand or foresee all the consequences (good or bad) that might come about from developing, deploying, or using new or emerging technologies. For this reason, new or emerging technologies could be misunderstood or have unforeseen consequences when humans, not God, are in charge of them.

Finally, as an omnibenevolent being, God's benevolence or moral goodness is maximal, which is why God is supposed to be the sort of being that would never misuse or abuse new or emerging technologies. In other words, any new or emerging technology, no matter how tempting it is to misuse or abuse it, is no match for God's perfect goodness. For this reason, new or emerging technologies could not be misused or abused when God is in charge of them. Unlike God, human beings are not all-good or morally perfect, and so humans could misuse or abuse new or emerging technologies. For this reason, new or emerging technologies could be misused or abused when humans, not God, are in charge of them.

It is important to note that the divine attributes, namely, omnipotence, omniscience, and omnibenevolence, map onto the fundamental questions of technology ethics discussed in Chapter 1. As we have seen in Chapter 1, the following questions express moral concerns that can arise with respect to many new or emerging technologies.

- Can we control it? (power, omnipotence)
- Do we understand it? (knowledge, omniscience)
- Will we be corrupted by it? (moral character/goodness, omnibenevolence)

These are fundamental questions about the limits of our power to control our environment, our knowledge of our world and our place in it, and our corruptible nature or moral character. The first question is about power and control. While an omnipotent being, which is what God is supposed to be, has unlimited power, which is why such a being could not lose control over anything, our power is limited, which is why losing control over new or emerging technologies is always a risk for us. The second question is about knowledge and understanding. While an omniscient being, which is what God is supposed to be, has unlimited knowledge, which is why such a being could not fail to understand anything, our knowledge is limited, which is why misunderstanding new or emerging technologies is always a risk for us. The third question is about moral character and virtue. While an omnibenevolent being, which is what God is supposed to be, is morally perfect, which is why such a being could not be tempted to misuse or abuse anything, our moral character is flawed, which is why misusing or abusing new or emerging technologies is always a risk for us.

Given this conception of God, then, humans "play God" when they pretend that they are as powerful as God is supposed to be, and so new or emerging technologies could not get out of control when they design, develop, deploy, or use them. Humans "play God" when they pretend that they are as prescient as God is supposed to be, and so new or emerging technologies could not be misunderstood or have unforeseen consequences when they design, develop, deploy, or use them. Humans "play God" when they pretend that they are as morally perfect as God is supposed to be, and so new or emerging technologies could not be misused or abused when they design, develop, deploy, or use them. Of course, humans are not as powerful, prescient, or morally perfect as God is supposed to be, and so new or emerging technologies could get out of control, be misunderstood or have unforeseen consequences, and be misused or abused by humans.[8] In that respect, the concept of God can be useful for "playing God" arguments precisely because it is the concept of an all-powerful, all-knowing, and all-good being.

I use the phrase "supposed to" when talking about the concept of God to indicate that our concern here is with the *concept* of God, not the *being*. That is, the Western theistic conception of God is that of a "maximally great (or perfect)" being (Hoffman and Rosenkrantz 2017). For present purposes, the real existence of a being like God is beside the point. That is, whether or not such a being exists, the concept is still that of a maximally great (or perfect) being just as, whether or not any triangles exist, the concept of a triangle is still that of a plane figure with three straight sides and three angles. For this reason, we can talk about the concept of God while putting aside the question about the existence of God. This point is important because it is quite common to object to "playing God" arguments by raising doubts about the existence of God. For example, according to Rocci Luppicini (2012, 235):

> [T]he fundamental problem with the "playing God" argument [. . .] is that *it requires us to grant the existence of a creator God*, who remains actively involved with his creation. This, however, is a leap of faith that many are unwilling to make (emphasis added).

My interpretation of "playing God" arguments, however, requires us to grant no such thing. Whether or not God exists, God is *supposed to be* all-powerful, all-knowing, and all-good *by definition*. This *concept* of a maximally great (or perfect) being, not the being itself, can be used to make arguments for or against the development or deployment of new or emerging technologies independently of the existence of the being denoted by that concept. So, there is no need to assume that God exists in order to make "playing God" arguments for or against new or emerging technologies any more than there is a need to assume that numbers exist in order to write mathematical proofs. The ontological question regarding the existence of numbers is a separate question, which is independent from the mathematical practice of writing proofs. Likewise, the ontological question regarding the existence of God is a separate question, which is independent from the discursive practice of making arguments. For this reason, "playing God" arguments cannot be dismissed simply by raising doubts about the existence of God.

Besides the ontological objection regarding the existence of God, to which my interpretation of "playing God" arguments is immune, Massimiliano Simons (2022) raises another objection against "playing God" arguments interpreted non-religiously as having no ontological commitment to the existence of God. Simons (2022, 155) argues that "playing God" arguments interpreted non-religiously, without ontological commitments to the existence of God, entail that the use of some new or emerging technology would be morally permissible "*if* the capacities of humans were more extensive (closer to an all-powerful, all-knowing, and all-good being)" (original emphasis). Now, if human beings were omnipotent, omniscient, and omnibenevolent, just as God is supposed to be, then they would *not* be "playing God." Recall that "playing God" means pretending that one is all-powerful, all-knowing, and all-good as God is supposed to be.[9] On my interpretation of "playing God" arguments, omnipotence, omniscience, and omnibenevolence are not "the appropriate properties to use [new or emerging] technologies" (Simons 2022, 155). Rather, they provide useful conceptual tools for deciding whether we *should* use some new or emerging technology. As Brian Green (2022, 146) puts my interpretation of "playing God" arguments, "the playing God argument is really *more of a critique of humanity than an appeal to anything supernatural*; it is an appeal to remember human frailty and to avoid hubris—and eventual humiliation—through the virtue of humility" (emphasis added).[10]

With this understanding of the role that the concept of God should play in "playing God" arguments in hand, we are now in a position to see what an argumentation scheme for "playing God" arguments, along with its Critical Questions, looks like. In the next chapter, then, I sketch such an argumentation scheme with its accompanying Critical Questions. In Chapter 4, I show how to analyze and evaluate "playing God" arguments using this argumentation scheme with its accompanying Critical Questions. Taken together, this "playing God" argumentation scheme and its accompanying Critical Questions constitute a general framework for thinking critically about the morality of new or emerging technologies, and thereby avoiding the extremes of techno-optimism and techno-pessimism. My hope is that this framework could help to inform approaches to the governance of new or emerging technologies.

Notes

1 Technology ethics is not the only domain in which "playing God" arguments can be made. For example, the debate over the morality of capital punishment features "playing God" arguments. According to Karl Menninger (1959, 15), for instance, capital punishment is morally wrong because it amounts to "playing God" (Mizrahi 2023).

2 On whether or not the phrase "playing God" has its origin in Christian theology, see Peters (2003). Peters argues that there is no place in Christian theology for the "playing God" argument. For Peters (2007, 182), the issue is not whether new or emerging technologies are morally objectionable because they interfere

with God's creation but rather whether emerging technologies "will respond to a transcendental ground for goodness and will enhance our capacity to love." On the other hand, concerns about "playing God" can be traced back to the Hebrew Bible for, insofar as God granted human beings dominion over nature, there is a danger that humans would abuse their authority over nature (see, e.g., Isaiah 45: 9–12). At any rate, secular "playing God" arguments can be, and indeed have been, advanced by non-religious philosophers. See, for example, Michael Sandel's (2007) "playing God" argument against human genetic enhancement.

3 I say more about that in Chapter 3.

4 It is important to note the difference between "playing God" arguments interpreted as warnings against interfering with God's creation or nature and so-called "appeals to nature." The latter are arguments that proceed from premises about what is natural or unnatural to conclusions about what is morally right or wrong. An example of such an argument is to argue that homosexuality is morally objectionable because it is unnatural (see, e.g., Corvino 2013, 78–9). Likewise, in the *History of Animals*, Aristotle says, "whatsoever is in conformity with nature is pleasant, and all animals pursue pleasure in keeping with their nature" (VIII.1, 598a7). And in *Politics*, Aristotle writes, "nothing which is contrary to nature is good" (VII.3, 1325b10). Thanks to an anonymous reviewer for this point.

5 It is also important to note the difference between the so-called "God complex" and the notion of "playing God." As far as I can tell, the former comes from psychoanalysis and the writings of the likes of Ernest Jones. According to Jones (1923, 207), "the main foundation of the [God] complex is to be discovered in a colossal *narcissism*, and this I regard as the most typical feature of the personalities in question. All the character-traits presently to be described can either be directly derived from narcissism, or else stand in the closest connection with it" (emphasis in original). Accordingly, unlike the "God complex," which has to do with a narcissistic personality, the notion of "playing God" pertains specifically to actions that can be evaluated from a moral point of view regardless of one's psychological makeup. For this reason, one can have a "God complex," that is, be a narcissist, without "playing God," although one might "play God," of course, and one can "play God" without being a narcissist (or having a "God complex").

6 "According to traditional western theism, classically expressed by Anselm and others, God is the greatest being possible" (Hoffman and Rosenkrantz 2013,

319). In philosophy of religion, "Western theism" generally refers to the so-called "Judeo-Christian tradition," that is, the God of the monotheistic religions: Judaism, Christianity, and Islam. For more on the latter and "playing God," see Shahzad (2007).

7 Of course, God would not *need* to use technology because God is supposed to be an omnipotent being. For this reason, it might seem a bit strange to say that a being like God would use technology. However, just as God is supposed to be all-powerful such that God would not need to use technology, God would also be able to use technology if God wanted to. In other words, as an all-powerful being, God could use technology if God wanted, even though God would not need to.

8 According to Paul Ramsey (1970, 138), "Men ought not to play God before they learn to be men, and after they have learned to be men they will not play God."

9 Could *not taking action* count as "playing God"? For example, if we have the technology to genetically modify mosquitoes such that they will not be able to infect humans with deadly diseases, but we choose not to use it, are we "playing God" in this case? As we have seen in Chapter 2, "playing God" is commonly understood as a form of doing or acting rather than allowing or omitting. As Peters (2007, 173) puts it, "By 'playing God', [. . .] [w]e mean changing nature." Accordingly, it seems that allowing nature to take its course would not count as "playing God."

10 "In doing so [i.e., deploying some new or emerging technology, such as gene editing], we risk making fundamental changes to both the world around us and to our very nature as human beings. Are we, therefore, 'playing God?' Is this a proper role for us? If it is, how do we exercise such tremendous power wisely, with humility and compassion?" (Herzfeld 2017, 1).

3

An Argumentation Scheme for "Playing God" Arguments

When the phrase "playing God" is used in debates concerning the morality of new or emerging technologies, such as cloning and genetic engineering, it is usually interpreted as a warning to refrain from interfering with God's creation or nature. But this interpretation of "playing God" arguments as a call to non-interference with nature may be too narrow. So, in Chapter 2, I put forth an alternative interpretation of "playing God" arguments. According to my interpretation of "playing God" arguments, first articulated in Mizrahi (2020), such arguments should be understood, not as warnings to leave God's creation or nature alone, but rather as invitations to think carefully about all the ways in which the development, deployment, or use of new or emerging technologies could go seriously morally wrong.

In this chapter, by taking an argumentation theory approach, I sketch an argumentation scheme and its accompanying Critical Questions that together capture the moral concerns expressed by "playing God" arguments. If successful, this "playing God" argumentation scheme and its accompanying Critical Questions together form a general framework for thinking critically about the morality of new or emerging technologies. The aim is that my argumentation scheme for "playing God" arguments and its accompanying Critical Questions can be useful conceptual tools for the analysis and evaluation of "playing God" arguments wherever they might occur as well as open new avenues for approaching the governance of new or emerging technologies.

According to Walton, Reed, and Macagno (2008, 1), "Argumentation schemes are forms of argument (structures of inference) that represent *structures of common types of arguments used in everyday discourse*, as well as in special contexts like those of legal argumentation and scientific argumentation" (emphasis added). As such, argumentation schemes are a useful tool for the analysis and evaluation of "playing God" arguments, since such arguments are commonly used in everyday discourse (Waytz and Young 2019). For instance, a 2018 Pew Research Center study shows that Americans often use "playing God" arguments when it comes to questions about biotechnology and genetic engineering. The results of this survey show that the main "reason for saying genetic engineering of mosquitoes to prevent the spread of some diseases by limiting their reproduction is taking technology too far" or that "genetic engineering to create more nutritious meat is taking technology too far" is that it amounts to "messing with nature and/or God's plan" (Funk and Hefferon 2018, 6), that is, "playing God."

Accordingly, "playing God" arguments are fairly common in everyday discourse. As Adam Waytz and Liane Young (2019, 1) observe, "This critique of playing God pervades people's opposition toward science and technology." Their "findings provide critical evidence for the association between aversion to playing God and moral condemnation of novel scientific practices," which include "robotics (drones), GMOs, vaccinations and stem cell research" (Waytz and Young 2019, 7).[1] But how should we analyze and evaluate "playing God" arguments? How do we know when a "playing God" argument is a good argument?

Here are two examples of "playing God" arguments from the aforementioned 2018 Pew Research Center study (emphasis added):

a) "Nature has selected species to become extinct over millions and millions of years. We have no right to bring animals back and *play God*" (quoted in Funk and Hefferon 2018, 9).
b) "[While] changing a fish to glow might sound like something people would want to see it's not something beneficial to humankind. At this point it would just [be] *playing God* to entertain rather [than] help us" (quoted in Funk and Hefferon 2018, 10).

As we have seen in Chapter 2, one way to understand these "playing God" arguments is as warnings to refrain from interfering with God's creation or nature. In both cases, the arguers are undoubtedly concerned about interfering with God's creation or nature. But it seems that these arguments are meant to express additional moral concerns. In other words, there is more to "playing God" arguments than just a concern about interfering with God's creation or nature. In particular, as we have seen in Chapter 2, "playing God" arguments invite us to think carefully about all the ways in which the design, development, deployment, or use of new or emerging technologies, such as genetic engineering, could go seriously morally wrong, given the fact that we are not as powerful, wise, and good as God is supposed to be.

For instance, in the first example above, (a), the arguer is suggesting that "bringing animals back" would be a bad idea, so we should not do it. Similarly, in the second example above, (b), the arguer is suggesting that using genetic engineering to make fish glow is not worth the risk, so it should not be done. Both arguers, then, seem to be concerned about how genetic engineering could go seriously wrong in ways that human beings, who are not as powerful, wise, and good as God is supposed to be, would fail to anticipate. As we have seen in Chapter 2, to "play God" is for humans to pretend that they are as powerful (all-powerful), wise (all-knowing), and benevolent (all-good) as God is supposed to be, and so bringing back extinct species or genetically modifying extant species by means of genetic engineering would not get out of control, be misunderstood or have unforeseen consequences, or be misused or abused by humans.

Now, like any logical argument, "playing God" arguments can be good or bad arguments. So, we need to be able to analyze and evaluate such arguments. That is, we need a way to capture the form and substance of such arguments so that we can tell whether they are good arguments whose conclusions we should accept or bad arguments whose conclusions we should reject. To this end, argumentation schemes are a useful tool. Accordingly, here is an argumentation scheme for "playing God" arguments that captures the aforementioned moral concerns about new or emerging technologies (cf. Mizrahi 2020, 1451).

Major Premise: We should not (morally speaking) "play God."
Minor Premise: Developing, deploying, or using technology *T* amounts to "playing God."
Conclusion: We should not (morally speaking) develop, deploy, or use *T*.

Applied to genetic engineering, for example, a "playing God" argument against using genetic engineering to bring back extinct species or genetically modify extant species would run as follows.

Major Premise: We should not (morally speaking) "play God."
Minor Premise: Using genetic engineering to bring back extinct species or genetically modify extant species amounts to "playing God."
Conclusion: We should not (morally speaking) use genetic engineering to bring back extinct species or genetically modify extant species.

Before the conclusion of this argument can be accepted or rejected, we should ask Critical Questions that are meant to "probe the weak points of an argument" (Walton, Reed, and Macagno 2008, 6). That is, an argumentation scheme is typically accompanied by a set of Critical Questions that together assist "in carrying out an analysis and evaluation of the argument" (Walton, Reed, and Macagno 2008, 11). As far as "playing God" arguments are concerned, the most crucial question is whether developing, deploying, or using technology, *T*, would indeed amount to "playing God."

As we have seen in Chapter 2, to "play God" is for humans (i) to pretend that they are as powerful as God is supposed to be, that is, all-powerful, and so new or emerging technologies would not get out of control if they develop, deploy, or use them, (ii) to pretend that they are as wise as God is supposed to be, that is, all-knowing, and so new or emerging technologies would not be misunderstood or have unforeseen consequences if they develop, deploy, or use them, and (iii) to pretend that they are as benevolent as God is supposed to be, that is, all-good, and so new or emerging technologies would not be misused or abused if they develop, deploy, or use them. To determine whether developing, deploying, or using a new or emerging technology counts as "playing God," then, we must determine whether it amounts to pretending that this new or emerging technology is manageable given limited human capacities as opposed to maximal divine attributes. In other words, the minor premise of the argumentation scheme for "playing God" argument is the key premise.

Accordingly, the following Critical Questions (CQ) are meant to accompany the aforementioned argumentation scheme of "playing God" arguments. These

Critical Questions are designed to help us determine whether developing, deploying, or using a new or emerging technology would count as "playing God" when a "playing God" argument is made (cf. Mizrahi 2020, 1452).

> CQ1: Could *T* get out of the control of (less than omnipotent) human beings?
> CQ2: Could *T* be misunderstood by or have unforeseen consequences for (less than omniscient) human beings?
> CQ3: Could *T* be misused or abused by (less than omnibenevolent) human beings?

These Critical Questions are conceptually related to the aforementioned scheme for "playing God" arguments in virtue of the concept of God invoked in such arguments. Unlike God, human beings can be overpowered, outwitted, or overwhelmed by the power, sophistication, and potential for abuse inherent in new or emerging technologies. Moreover, CQ1–CQ3 map onto the fundamental questions at the core of many debates over the morality of new or emerging technologies. These questions are the following (see Chapter 1 and Chapter 2).

- Can we control it? (omnipotence; CQ1)
- Do we understand it? (omniscience; CQ2)
- Will we be corrupted by it? (omnibenevolence; CQ3)

As we have seen in Chapters 1 and 2, these are fundamental questions about the limits of our power to control our environment, our knowledge of our world and our place in it, and our corruptible nature or moral character.

CQ1 covers what Sandler (2014, 10) calls a "power analysis." According to Sandler (2014, 10), "A prominent theme in the ethics of emerging technologies is *the need to take responsibility for the power that modern technology provides* and to develop ethics appropriate for that power" (emphasis added). As we have seen in Chapter 1, advancements in science and technology enhance our power to act in ways we were unable to act before and allow us to make choices we were unable to make before. When we think about the powers that new or emerging technologies bestow on us, we should think about whether we would be able to keep those new powers under our control. For instance, old space artifacts, such as obsolete spacecrafts or defunct satellites, become space debris (or space junk) over which we have very little control (Pelton 2013). So, when it comes to new or emerging technologies, we should think about the following questions: Who will control the technology? How will the technology be controlled? Who will have access to the technology? Who will not have access to the technology? Who will be empowered by it? Who will be disempowered by it? Who will be manipulated by the technology against their will or without their consent? Who will benefit from it? Who will not benefit from it? Such questions, and others like them, are covered by CQ1, which stems from the fundamental question about our limited power (i.e., Can we control it?) in comparison with the maximal power of an all-powerful (omnipotent) being.

CQ2 covers what Sandler (2014, 12) calls "Environment, Health, and Safety (EHS) concerns," which have "to do with the *possible negative impacts of technology* on human welfare and the nonhuman environment" (emphasis added). When we think about the opportunities that new or emerging technologies afford us, we should also think about whether we truly understand those opportunities and their implications. For, if they are misunderstood, opportunities may easily turn into perils, risks, or threats. After all, as Sandler (2014, 12) observes, the "negative EHS impacts of an emerging technology are typically *unintended* and *unwanted*" (emphasis added). In other words, new or emerging technologies tend to be disruptive. Disruptive technologies, by their very nature, could be misunderstood by, and have unforeseen consequences for, less than all-knowing beings like us. According to John Danaher and Jeroen Hopster (2022, 10), "Various other potentially disruptive technologies, such as autonomous weapon systems, solar radiation management, or the artificial womb, may similarly give rise to harmful uncertainty in defined domains of inquiry."[2] For instance, while advancements in food processing technology have helped to provide a stable supply of food to more people around the world than ever before, it has become apparent that we have seriously misunderstood the potential harmful effects of processed food on human health (Mellor 2024). So, when it comes to new or emerging technologies, we should think about the following questions: Do we fully understand the opportunities presented by the technology? Do we fully understand the challenges presented by the technology? What are the potential risks? What are the possible threats? Such questions, and others like them, are covered by CQ2, which stems from the fundamental question about our limited knowledge (i.e., Do we understand it?) in comparison with the maximal knowledge of an all-knowing being.

CQ3 covers what Sandler (2014, 14) calls the "dual use challenge of emerging technologies." For, in addition to *unintended impacts*, new or emerging technologies tend to have *unintended uses* as well. As Sandler (2014, 14) explains, "Many technologies are *dual use* technologies in that they can be effectively employed in ways other than their intended or designed use" (original emphasis). When we think about the powers and opportunities that new or emerging technologies grant us, we should also think about whether we have the moral character to use new or emerging technologies for good and the moral fortitude to resist misusing or abusing such technologies, no matter how tempting it may be to do so. After all, as Sandler (2014, 2) observes, "*Technology shapes us and our social and ecological world as much as we shape technology*" (original emphasis). For instance, automobiles have changed not only our social environments, for example, our cities, but also how we treat one another, for example, road rage (Baer 2019). So, when it comes to new or emerging technologies, we should think about the following questions: How will we use the technology? Can it be easily misused? Can it lead to abuse? Will it make us dependent on it? How will it impact our relationships with others? Such questions, and others like them, are covered by CQ3, which stems from the fundamental question about our limited goodness (i.e., Will we be corrupted by it?) in comparison with the maximal goodness of an all-good being.

Now, an evaluation of a "playing God" argument will depend on answers to Critical Questions CQ1–CQ3. More specifically, affirmative answers to CQ1–CQ3 provide good reasons to think that developing, deploying, or using new or emerging technology *T* would indeed amount to "playing God" in terms of power, knowledge, and benevolence. That is, if we were to use *T*, we would thereby be pretending that we could not lose control over *T*, misunderstand or be caught off guard by the unforeseen consequences of developing, deploying, or using *T*, or be tempted to misuse or abuse *T*. This, in turn, provides a good reason to think that we should not use *T* until we have a better understanding of all the ways in which developing, deploying, or using

T could go seriously morally wrong. On the other hand, negative answers to CQ1–CQ3 provide good reasons to think that developing, deploying, or using new or emerging technology *T* would not amount to "playing God." That is, if we were to use *T*, we would not be pretending that we could not lose control over *T*, misunderstand or be caught off guard by the unforeseen consequences of developing, deploying, or using *T*, or be tempted to misuse or abuse *T*.

On my construal, "playing God" arguments are defeasible arguments. Defeasible reasoning is not deductive reasoning. This means that the premises of a good defeasible argument, if true, provide some evidential support for the conclusion, but they do not guarantee the truth of the conclusion. Accordingly, if the major and minor premises of a "playing God" argument are true, the conclusion of that argument would then be supported by the evidence cited in those premises, but the conclusion can always be defeated by additional evidence that is not cited in the premises. Now, given that defeasible arguments can be stronger or weaker, my argumentation scheme for "playing God" arguments and its associated Critical Questions, CQ1–CQ3, provide a straightforward way of telling how strong a "playing God" argument is. That is, the more Critical Questions are answered in the affirmative, the stronger the argument, whereas the more Critical Questions are answered in the negative, the weaker the argument.

To illustrate using the example of genetic engineering again, we would ask the following Critical Questions in order to determine whether we should accept or reject the conclusion of the aforementioned "playing God" argument according to which we should not use genetic engineering to bring back extinct species or genetically modify extant species because doing so amounts to "playing God."

CQ1: Could the genetic engineering of animals get out of the control of (less than omnipotent) human beings?

CQ2: Could the genetic engineering of animals be misunderstood or have unforeseen consequences for (less than omniscient) human beings?

CQ3: Could the genetic engineering of animals be misused or abused by (less than omnibenevolent) human beings?

If the answer to all of these Critical Questions is "yes," then it means that using genetic engineering to bring back extinct species or genetically modify extant species would indeed amount to "playing God" in terms of power, knowledge, and benevolence. Given that we should not pretend to be all-powerful, all-knowing, and all-good as God is supposed to be, we could then conclude that we should not use genetic engineering to bring back extinct species or genetically modify extant species until we have a better understanding of all the ways in which genetic engineering could go seriously morally wrong. In this case, we would have a strong "playing God" argument against using genetic engineering to bring back extinct species or genetically modify extant species.

On the other hand, if the answer to all of these Critical Questions is "no," then it means that using genetic engineering to bring back extinct species or genetically modify extant species would not amount to "playing God." In that case, the conclusion that we should not use genetic engineering to bring back extinct species or genetically modify extant species would not be warranted. In that case, we would have a weak "playing God" argument against using genetic engineering to bring back extinct species or genetically modify extant species.[3]

Any other combination of "yes" and "no" answers to the Critical Questions, CQ1–CQ3, would make the "playing God" argument stronger or weaker. That is, the more Critical Questions are answered in the affirmative, the stronger the argument, whereas the more Critical Questions are answered in the negative, the weaker the argument. Table 3.1 lays out the possible combinations of "yes" and "no" answers to the Critical Questions, CQ1–CQ3, and the corresponding strength of a "playing God" argument.

Our answers to the Critical Questions will also tell us what issues need to be addressed before the new or emerging technology in question can be deployed or used. For example, if the answer to CQ1 and CQ2 is "no," but the answer to CQ3 is "yes," then we would have a somewhat weak "playing God" argument against using genetic engineering to bring back extinct species or genetically modify extant species. Nevertheless, we would have reason to believe that, as far as genetic engineering technology is concerned, the main moral concern is not that it could get out of control, be misunderstood, or have unintended, negative consequences. Rather, the main moral concern is that it could be misused or abused by human beings who, unlike a supposedly omnibenevolent

Table 3.1 The Strength of "Playing God" Arguments as a Measure of Possible Answers to Critical Questions, CQ1–CQ3, of the Argumentation Scheme for "Playing God" Arguments

CQ1	CQ2	CQ3	Strength of "Playing God" Argument
yes	yes	yes	strong (power, knowledge, benevolence)
yes	yes	no	somewhat strong (power, knowledge)
yes	no	yes	somewhat strong (power, benevolence)
no	yes	yes	somewhat strong (knowledge, benevolence)
yes	no	no	somewhat weak (power)
no	yes	no	somewhat weak (knowledge)
no	no	yes	somewhat weak (benevolence)
no	no	no	weak

Source: Author

being, are not morally perfect or all-good. Having such a reason would help us to direct our governance efforts toward crafting policies, regulations, or laws designed to prevent misuse or abuse instead of more research (to address CQ2) and safety testing (to address CQ1).[4]

In *Fluke* (2024), the political scientist Brian Klaas refers to the evolutionary biologist Richard Lenski's *E. coli* long-term evolution experiment as "playing God." In particular, Klaas invites us to imagine "the ability to play God, pressing pause at will, even rewinding and replaying key moments [of evolutionary history]" (Klaas 2024, 51). This raises an interesting question: Could scientific experimentation count as "playing God"? For example, if scientists use technology to genetically modify animals such that they glow (e.g., using Green Fluorescent Protein or GFP) for the purpose of scientific experimentation, would the scientists be "playing God" in this case? I think that this case is covered by the argumentation scheme for "playing God" arguments and its accompanying Critical Questions. That is, scientists who are considering using GFP or other so-called "glowing genes" (Zimmer 2005) in their experiments with animals should ask themselves the following Critical Questions.

CQ1: Could using GFP in scientific experimentation get out of the control of (less than omnipotent) human beings?

CQ2: Could using GFP in scientific experimentation be misunderstood or have unforeseen consequences for (less than omniscient) human beings?

CQ3: Could using GFP in scientific experimentation be misused or abused by (less than omnibenevolent) human beings?

If they answer all of these Critical Questions in the affirmative, then the scientists would have good reasons to think that using GFP in scientific experimentation would indeed amount to "playing God" in terms of power, knowledge, and benevolence. On the other hand, if they answer all of these Critical Questions in the negative, then the scientists would have good reasons to think that using GFP in scientific experimentation would not amount to "playing God." In general, the more Critical Questions they answer in the affirmative, the stronger the "playing God" argument against using GFP in scientific experimentation on animals would be. On the other hand, the more Critical Questions they answer in the negative, the weaker the "playing God" argument against using GFP in scientific experimentation on animals would be. If the scientists answer all of these Critical Questions in the negative, then they would have no good reason to think that using GFP in scientific experimentation amounts to "playing God." This is how the argumentation scheme for "playing God" arguments and its associated Critical Questions, CQ1–CQ3, provide a rather straightforward way of telling how strong (or weak) a "playing God" argument is.

As Walton, Reed, and Macagno (2008, 20) point out, we can "add the content of each of the Critical Questions as a separate premise," if we prefer to see the argument laid out in its entirety. As far as the aforementioned argumentation scheme for "playing God" arguments is concerned, doing so results in the following extended argumentation scheme (cf. Mizrahi 2020, 1454).

Major Premise: We should not (morally speaking) "play God."
Conditional Premise: If *T* could get out of the control of humans, be misunderstood or have unforeseen consequences for humans, or be misused or abused by humans, then developing, deploying, or using *T* amounts to "playing God."
Control Premise: *T* could get out of the control of humans.
Understanding Premise: *T* could be misunderstood or have unforeseen consequences for humans.
Use Premise: *T* could be misused or abused by humans.
Conclusion: We should not (morally speaking) develop, deploy, or use *T*.

In this extended argumentation scheme for "playing God" arguments, the *Control Premise* corresponds to CQ1, the *Understanding Premise* corresponds to CQ2, and the *Use Premise* corresponds to CQ3.

An advantage of my interpretation of "playing God" arguments, encapsulated in the argumentation scheme for "playing God" arguments and its accompanying Critical Questions, CQ1–CQ3, outlined above, is that such arguments do not commit us to the existence of God or to any views about the creation of the universe. For humans' limitations in power, knowledge, and goodness in comparison to a being like God hold whether or not God exists and whether or not the universe was created by an "omniscient, omnipotent, omnibenevolent, eternal creator" (Oppy 2006, 16). For example, arguing that "HGE [i.e., Human Genetic Engineering] is an arrogant interference with the nature that God has created" (Evans 2002, 125), and hence morally wrong, seems to commit the arguer to the existence of God as the creator of nature. Consequently, the conclusion of this "playing God" argument, that is, that human genetic engineering is morally impermissible because HGE interferes with God's creation, can be rejected merely on the grounds that there is no compelling evidence for the existence of God. As Gordon Graham (2002, 145) puts it, "if, as the secular world believes, there *is* no God, how could there be any danger of human beings illegitimately abrogating to themselves His function?"

On my interpretation of "playing God" arguments, however, there is no need to postulate the existence of God as the creator of the universe. Human limitations hold irrespective of the existence of God. As beings that are limited in power, knowledge, and goodness, humans should always consider carefully all the ways in which the development, deployment, or use of new or emerging technologies could go seriously morally wrong before applying those technologies. For this reason, even non-believers (i.e., "the secular world") can make "playing God" arguments, on my interpretation of "playing God" arguments, for it allows us to remain non-committal on the question concerning God's existence. In other words, theists, atheists, and agnostics of various stripes can be morally concerned about all the ways in which the application of new or emerging technologies could go seriously morally wrong at the hands of beings that are limited in power, knowledge, and goodness in comparison to a "maximally great (or perfect)" being (Hoffman and Rosenkrantz 2017). Thus understood as inviting us to think carefully about all the ways in which the development, deployment, or use of new or emerging technologies could go seriously morally wrong, not as warning us to leave God's creation or nature alone, "playing God" arguments cannot be dismissed simply on the grounds that there is no compelling evidence for the existence of God.

In addition to being neutral on questions about God's existence and creation, another advantage of my interpretation of "playing God" arguments is that such arguments can be made even when the new or emerging technology in question is not designed to intervene in nature directly. That is, on my interpretation, "playing God" arguments can be advanced in domains other than bioethics, where the technology in question is designed to intervene in human nature specifically. To illustrate this point, consider the case of social media, that is, internet-based websites and applications (or apps), such as Facebook, YouTube, Snapchat, and TikTok, which enable users to engage in social networking. Unlike cloning, genetic engineering, and other technologies in the domain of biomedical science and technology, social media technologies are not designed to intervene in nature directly, so they fall outside of the domain of biomedical ethics. Nonetheless, I think that technological advances in social media raise many of the moral concerns that can be captured by the "playing God" argumentation scheme and its accompanying Critical Questions.

Indeed, Jake Wallis Simons has made a "playing God" argument against one social media platform in particular, namely, Snapchat, in an article published in *The Spectator*, titled "Snapchat Is Playing God with Our Children's Wellbeing" (2023). Simons (2023) writes:

> [T]he introduction of an *experimental chatbot* onto the phones of teenagers across the world, without a warning or even permission, is deeply disturbing in itself. It's no exaggeration to suggest that *Snapchat is playing God with the wellbeing of our kids*. But how can parents fight back? Attempts at curbing the power of Silicon Valley are as ineffective as attempts to keep teenagers away from their phones. A worldwide boycott of Snapchat is hardly realistic (emphasis added).

Accordingly, Simons (2023) is expressing moral concerns about the ways in which social media companies experiment with chatbots on their platforms, more specifically, Snapchat, given that the consequences of such experiments in real time for the well-being and health of young users are not very well understood yet. This concern is related directly to one of the fundamental questions about the limits of our knowledge. As we have seen in Chapter 1, the following questions express moral concerns that can arise with respect to many new or emerging technologies, including social media technologies.

- Can we control it?
- Do we understand it?
- Will we be corrupted by it?

These are fundamental questions about the limits of our power to control our environment, our knowledge of our world and our place in it, and our corruptible nature or moral character. As such, they map onto the Critical Questions of the argumentation scheme for "playing God" arguments.

CQ1: Could *T* get out of the control of (less than omnipotent) human beings?
CQ2: Could *T* be misunderstood by or have unforeseen consequences for (less than omniscient) human beings?
CQ3: Could *T* be misused or abused by (less than omnibenevolent) human beings?

So, to see how the "playing God" argumentation scheme and its accompanying Critical Questions apply to technologies beyond biomedical technologies, such as digital technologies like social media, which are not designed to intervene in nature directly, we need to ask Critical Questions, CQ1–CQ3, with respect to those technologies. Let's start with the first Critical Question, CQ1, then. Could social media get out of the control of (less than omnipotent) human beings?

When it comes to using social media websites and apps, whether these technologies could get out of control is a legitimate moral concern. Arguably, social media can be said to be getting out of control already in at least two respects. First, there is hardly any control over what gets posted or shared on social media platforms and by whom. This means that a lot of the content posted and shared on social media is untrustworthy and even harmful. Prominent examples of disinformation and misinformation on social media include the Internet Research Agency's efforts to undermine American democracy and influence the 2016 US presidential election (Eady et al. 2023) as well as the anti-vaccination movement's efforts to spread misinformation about COVID-19 vaccines (Lee et al. 2022). The introduction of chatbots to social media platforms could accelerate the spread of misinformation because these technologies can be used to generate more misinformation faster.[5]

It is important to note that the spread of misinformation and disinformation on social media is not an instance of users misusing or abusing social media websites or apps. Rather, social media platforms seem to be designed in such a way that misinformation, disinformation, fake news, and the like spread wider and faster than accurate information. That is, the recommendation algorithms that drive social media seem to push misinformation and disinformation to the front of the line, so to speak, at the expense of accurate information (Vosoughi, Roy, and Aral 2018). In that sense, then, social media can be said to be getting out of control insofar as misinformation, disinformation, fake news, and the like pervade social media platforms despite our best interests, intentions, or efforts.

The second way in which social media can be said to be getting out of our control is in terms of compulsive use or excessive consumption of social media. As Cecilia Cheng et al. (2021) put it:

> [P]roblematic use [of social media] can impair users' psychosocial functioning and well-being. [. . .] For instance, some individuals are so engaged in Instagram that they feel distressed when they are unable to use it during work. Such misuse is widely referred to as *social media addiction* (emphasis added).

Even though the researchers talk about misuse here, we should be careful about blaming the victims of social media addiction. For, in a sense, they are using social media precisely as intended. That is, social media technologies are designed to keep users engaged and on the websites or apps for as long as possible. After all, infinite scroll or endless scrolling (e.g., Instagram), autoplay (e.g., YouTube), streaks (e.g., Snapchat), and the like are not bugs. They are design features whose aim is to keep users on social media websites and apps longer than they would otherwise be. Indeed, several US states and the District of Columbia have filed a lawsuit against Meta, the parent company of Facebook, Instagram, and WhatsApp, which alleges that Meta has deliberately engineered its social media platforms to be addictive to young users (Ortutay 2023). A similar lawsuit, which alleges violating consumer protection laws and contributing to "a teen mental health crisis" as the app is "deliberately designed to keep young people hooked on the service" (Allyn 2024), was filed against TikTok. Now, "Addiction is [. . .] characterized by inability to constantly abstain, diminished behavioral control, craving, diminished recognition of significant problems with one's behavior and dysfunctional emotional response" (Ndasauka et al. 2017, 4). So, social media platforms seem to be designed to make us unable to abstain from using them, to diminish our control over using them, and to make us crave using them while failing to recognize that our usage of social media has become or is becoming problematic.

In fact, recently revealed internal documents from TikTok show that the company knew that its app is causing harm to its users (Allyn, Goodman, and Kerr 2024). According to these documents, TikTok concluded from its own research that "compulsive usage correlates with a slew of negative mental health effects like loss of analytical skills, memory formation, contextual thinking, conversational depth, empathy, and increased anxiety" (Allyn, Goodman, and Kerr 2024). TikTok was also aware that "compulsive usage also interferes with essential personal responsibilities like sufficient sleep, work/school responsibilities, and connecting with loved ones" (Allyn, Goodman, and Kerr 2024). Additionally, the company's own research determined "the precise amount of viewing it takes for someone to form a habit: 260 videos. After that [. . .] a user 'is likely to become addicted to the platform'" (Allyn, Goodman, and Kerr 2024). Given that TikTok videos tend to be short, even just seconds long, and are curated for viewers in quick succession algorithmically, this means that "in under 35 minutes, an average user is likely to become addicted to the platform" (Allyn, Goodman, and Kerr 2024). In that sense, social media can be said to be getting out of control insofar as social media addiction is a growing problem.

If social media technologies are already getting out of control in the two ways described above, namely, in terms of loss of control over the misinformation and disinformation spreading on social media platforms and in terms of users' loss of control over habitual use of social media websites or apps, then the answer to CQ1 seems to be "yes." That is, social media use and consumption could get out of the control of (less than omnipotent) human beings. While God, as a supposedly omnipotent being, would never lose control over social media, we are not all-powerful. As we have seen, it appears that we have already lost control over social media in at least two respects.

Let's move on to the second Critical Question, CQ2. Could social media be misunderstood by or have unforeseen consequences for (less than omniscient) human beings? Arguably, we still do not fully understand social media and its effects on us, especially the young among us. Only recently, we discovered that social media can be addictive, but we still do not understand why that is the case. Yalin Sun and Yan Zhang (2021) identified no less than twenty-five theories and models that guide research on social media addiction. We are also somewhat in the dark about the ways in which social media affects the mental health of users, especially young users. We know that the "rapid growth of SM [Social Media] in the lives of youth has seen a parallel rise in depression and anxiety, both reaching unprecedented heights by 2020" (Weigle and Shafi 2024). This association between social media use and teen mental health was enough for the US Surgeon General, Dr. Vivek Murthy, to issue an advisory about the effects of social media use on youth mental health. But we do not yet know if social media use *causes* depression and anxiety among young people. As the US Surgeon General admits in the new advisory about social media use and youth mental health, "we do not have enough evidence to conclude that it [i.e., social media] is sufficiently safe for" teenagers, which is why "more research is necessary" (The US Surgeon General's Advisory 2023, 10–11).

If we already fail to fully understand social media and its potentially harmful effects on us and our mental health, then the answer to CQ2 seems to be "yes." That is, social media could be misunderstood by or have unforeseen consequences for (less than omniscient) human beings. While God, as a supposedly omniscient being, would never misunderstand or fail to anticipate the harm that might be inherent in using or consuming social media, we are not all-knowing. As we have seen, it appears that we have already misunderstood or failed to appreciate the potentially harmful effects that social media use could have on us.

Let's move on to the third Critical Question, CQ3. Could social media be misused or abused by (less than omnibenevolent) human beings? The promise of social media, such as Facebook, was to make it easier for people to connect with others no matter where they are in the world. For example, on the homepage of Facebook, the tagline reads, "Connect with friends and the world around you on Facebook" (www.facebook.com). But the perils of social media are many. We have already seen that social media can be, and has been, used to spread misinformation, disinformation, fake news, and the like. Social media can also be used to commit fraud and other crimes. The Federal Trade Commission (FTC) calls social media "a golden goose for scammers" (Fletcher 2023). In fact, nowadays, scams often start on social media. According to an FTC report, "One in four people who reported losing money to fraud since 2021 said it started on social media" (Fletcher 2023). Social media scams have cost Americans roughly 2.7 billion dollars from January 2021 through June 2023, which is probably "just a small fraction of the public harm," since "the vast majority of frauds are not reported" (Fletcher 2023).

For instance, a particularly cruel form of social media scams is known as romance scams. Romance scammers prey on people who seek companionship, friendship, or love on social media platforms or dating apps. Scammers contact victims through social media websites or apps, such as Instagram or Facebook. They chat with their victims and begin to develop a long-term online relationship with them. When they feel that enough trust has been built, the scammers begin to ask the victims for money, often in the form of gift cards, wire transfers, cryptocurrency, and the like. According to an FTC report, "In the first six months of 2023, half of people who said they lost money to an online romance scam said it began on Facebook, Instagram, or Snapchat" (Fletcher 2023). Clearly, social media scams, like romance scams, which harm social media users financially and psychologically, are a far cry from the promise of social media as a set of technologies designed to connect people.

Arguably, one of the most egregious abuses of social media is the abuse perpetrated by digital tech companies themselves. The tech companies that make, own, and operate social media websites and apps use them to collect user data, often without the explicit, voluntary, and informed consent of users. Personal user data are then used to inundate social media users with targeted advertisements, thereby disregarding user privacy. For instance, in a now infamous case of user data breach, the now defunct consulting firm Cambridge Analytica harvested the personal data of millions of Facebook users for the purposes of voter profiling and political targeting. An investigation by the UK's Information Commissioner's Office (ICO) "found that Facebook breached data protection laws by failing to keep users' personal information secure, allowing Cambridge Analytica to harvest the data of up to 87 million people without their consent" (Zialcita 2019). Clearly, targeted advertising and voter profiling, which violate the privacy of social media users, are a far cry from the promise of social media as a set of technologies designed to connect people.

If social media technologies are already being misused and abused in many ways not only by users but also by the very tech companies that make, own, and operate social media websites and apps, then the answer to CQ3 seems to be "yes." That is, social media could be misused or abused by (less than omnibenevolent) human beings. While God, as a supposedly omnibenevolent being, would never be tempted to misuse or abuse social media, we are not all-good. In that respect, it seems clear that social media technologies are already changing us, but apparently not for the better. For, as we have seen, it appears that we have already misused and abused social media in several ways.

When it comes to social media, Critical Questions, CQ1–CQ3, seem to have affirmative answers. This means that developing, deploying, and using social media platforms would indeed amount to "playing God" in terms of power (CQ1), knowledge (CQ2), and benevolence (CQ3). Therefore, a "playing God" argument against social media is a strong argument (see Table 3.1). This argument can be summed up as follows.

Major Premise: We should not (morally speaking) "play God."

Conditional Premise: If social media could get out of the control of humans, be misunderstood or have unforeseen consequences for humans, or be misused or abused by humans, then developing, deploying, or using social media amounts to "playing God."

Control Premise: Social media could get out of the control of humans.

Understanding Premise: Social media could be misunderstood or have unforeseen consequences for humans.

Use Premise: Social media could be misused or abused by humans.

Conclusion: We should not (morally speaking) develop, deploy, or use social media.

This strong "playing God" argument against social media points to several moral problems with social media technologies, namely, that they could get out of control, that they could be misunderstood, that they could have unforeseen consequences, and that they could be misused or abused.

Like all "playing God" arguments, this one against social media is a defeasible argument. Conclusions of defeasible arguments can always be defeated by additional evidence that is not cited in their premises. In other words, this argument is not the final word about whether or how we should use social media. One obvious way to defeat this "playing God" argument against social media is to argue that the potential benefits of social media technologies would outweigh the perils of loss of control, misunderstanding, unforeseen consequences, misuse, and abuse. In other words, various ways to assess risks (e.g., in terms of probability of harm) might tip the scales either for or against the development and deployment of social media technologies.[6] Still, this strong "playing God" argument against social media gives us pretty good reasons to be concerned about these technologies, at least until we can make sure that we are able to control them, understand them and their effects, and use them benevolently.

Against the argumentation scheme for "playing God" arguments and its associated Critical Questions, it might be argued that there is no need for an argumentation scheme for "playing God" arguments in particular. Instead, it might be argued, the moral concerns expressed by "playing God" arguments can be captured adequately by existing argumentation schemes, especially the argumentation scheme for arguments from negative consequences. Walton, Reed, and Macagno (2008, 101) provide the following argumentation scheme for arguments from consequences, particularly *negative* consequences:

If I (an agent) bring about (don't bring about) *A*, then *B* will occur.
B is a bad outcome (from the point of view of my goals).[7]
Therefore, I should not (practically speaking) bring about *A*.[8]

However, there are crucial differences between Walton, Reed and Macagno's argumentation scheme for arguments from negative consequences and my argumentation scheme for "playing God" arguments. First, according to Walton, Reed, and Macagno (2008, 100), "Argument from consequences may be considered as a generic category of causal inference." Accordingly, the major premise of their argumentation scheme for arguments from negative consequences is about a causal relation between two things, *A* and *B*. In contrast, my argumentation scheme for "playing God" arguments does not fall under the category of causal inference. Accordingly, there is no premise about a causal relation between two things, *A* and *B*, in my argumentation scheme for "playing God" arguments. As we have seen in Chapter 2, what is doing the theoretical work in my interpretation of "playing God" arguments is the divine attributes that God is supposed to have by definition, namely, omnipotence, omniscience, and omnibenevolence. Instead of causal relations, then, my argumentation scheme for "playing God" arguments is about the conceptual relation between the concept of an omnipotent, omniscient, and omnibenevolent being, that is, God, in comparison with human beings who are not all-powerful, all-knowing, or all-good.

Second, according to Walton, Reed, and Macagno (2008, 101), arguments from consequences are also instances of practical reasoning, which "is often a backward process from the outcome of a consequence to its premises." Accordingly, the conclusion of their argumentation scheme for arguments from negative consequences is about what should not be brought about, *practically speaking*. In contrast, my argumentation scheme for "playing God" arguments is not an instance of practical reasoning. Accordingly, the conclusion of my argumentation scheme for "playing God" arguments is about what should (or should not) be done *morally speaking*, not practically speaking. The moral and the practical are often distinct. In other words, there may be practical reasons for doing something that should still not be done *morally speaking*. For example, if I am one of two survivors of an airplane crash who are stranded on a desert island with no food, I may have a practical reason to kill and eat the other survivor, given that my goal is to survive, even though I should not do it, morally speaking.[9]

Given these crucial differences between the argumentation scheme for "playing God" arguments and Walton, Reed, and Macagno's (2008) argumentation scheme for arguments from negative consequences, there seems to be a need for an argumentation scheme for "playing God" arguments that is different from the argumentation scheme for arguments from negative consequences.[10] My argumentation scheme for "playing God" arguments is sufficiently general to be able to cover "playing God" arguments that can be made in various domains, from biomedical science to digital technology, where new technologies emerge. In other words, my argumentation scheme for "playing God" arguments and its accompanying Critical Questions are meant to be conceptual tools for the analysis and evaluation of "playing God" arguments wherever they might occur. Taken together, my argumentation scheme for "playing God" arguments and its accompanying Critical Questions can provide a constructive lens through which to determine whether we need to make sure that we are able to control, understand, and benevolently use a new or emerging technology before it is morally permissible for us to deploy and use it.

In the next chapter, I show how my argumentation scheme for "playing God" arguments and its accompanying Critical Questions can be used to analyze and evaluate "playing God" arguments. Chapter 4 concludes Part I of this book, which provides a general theoretical framework for thinking critically about the morality of new or emerging technologies rather than falling into the opposite traps of techno-optimism on the one hand and techno-pessimism on the other hand. The aim is that my argumentation scheme for "playing God" arguments and its accompanying Critical Questions can be useful conceptual tools for the analysis and evaluation of "playing God" arguments wherever they might occur as well as open new avenues for approaching the governance of new or emerging technologies. In Chapter 4, I use familiar examples of biomedical technologies, such as Assisted Reproductive Technologies (ART) like in vitro fertilization (IVF), to show that my argumentation scheme for "playing God" arguments and its accompanying Critical Questions can handle such examples. Then, in Part II of this book, I analyze and evaluate four "playing God" arguments pertaining to new or emerging technologies beyond the biomedical domain as use cases for the analysis and evaluation of "playing God" arguments using my argumentation scheme for "playing God" arguments and its accompanying Critical Questions. These use cases include the following: climate engineering or geoengineering (Chapter 5), Artificial Intelligence or AI (Chapter 6), autonomous vehicles or self-driving cars (Chapter 7), and Lethal Autonomous Weapon Systems or LAWS (Chapter 8). Each chapter is meant to be a standalone use case that can be used as a case study in an applied ethics course, such as Technology Ethics, Engineering Ethics, AI Ethics, and the like.

Notes

1 For a "playing God" argument with respect to human mood enhancement technologies, see Kamenjasevic (2024). On "playing God" and synthetic biology, see van den Belt (2009).

2 On solar radiation management, see Chapter 5. On Lethal Autonomous Weapon Systems (LAWS), see Chapter 8.

3 Of course, "playing God" arguments are only one kind of argument being made in the debate concerning the morality of genetic engineering. There are plenty of other arguments having to do with justice, fairness, equality, human well-being, human rights, and many other issues pertinent to the debate. See, for example, Roberta Berry (2007) for more on such arguments. In that sense, too, "playing God" arguments are defeasible insofar as their conclusions can be defeated by additional evidence.

4 I say more about that in Chapter 4 where I discuss how answers to the Critical Questions, CQ1–CQ3, could help to inform approaches to technology governance.

5 See Chapter 6 on the use of Generative AI (GenAI) to create synthetic media, such as deepfakes.

6 I say more about this when I discuss how my argumentation scheme for "playing God" arguments and its associated Critical Questions may help to inform approaches to technology governance in Chapter 4.

7 In Walton, Reed, and Macagno's (2008) argumentation scheme for arguments from negative consequences, it is implicitly assumed, but not explicitly stated, that one should not bring about outcomes that are bad from the point of view of one's goals.

8 I follow Walton, Reed, and Macagno (2008) in using "practical reasoning" to talk about reasoning with respect to actions, not statements. According to Walton, Reed, and Macagno (2008, 94–5), "A practical inference has two characteristic types of premises. One states that the rational agent has a particular goal. The other states some kind of action through which the agent could carry out the goal." Note the use of "practically speaking" in their argument scheme for arguments from consequences quoted above.

9 It is important to note that the moral and the practical are not always so easily distinguishable. This is especially the case if one thinks of morality as action-guiding. In moral philosophy, the idea that morality is supposed to be action-guiding is often encapsulated in the principle known as "ought implies can." The idea is that one ought to do something only if one can do it. Conversely, if one cannot do something, then one has no moral obligation to do it. For example, if I ought to keep my promise to meet a friend for dinner, then I can keep my promise to meet a friend for dinner. But if I cannot meet a friend for dinner, then I have no moral obligation to do so. The "ought implies can" principle has been challenged by empirical evidence suggesting that people generally do assign moral obligations to agents who are incapable of fulfilling those obligations. See, for example, Kissinger-Knox, Aragon, and Mizrahi (2018) and Wesley Buckwalter (2019) for recent discussions.

10 For the same reasons, Walton, Reed, and Macagno's argumentation schemes for fear and danger appeals (Walton, Reed, and Macagno 2008, 102–4) are different from the argumentation scheme for "playing God" arguments, and thus fail to capture the moral concerns expressed by "playing God" arguments.

4

How to Analyze and Evaluate "Playing God" Arguments

On February 21, 2024, the state of Alabama's Supreme Court ruled that frozen embryos stored in fertility clinics are children, which means that destroying frozen embryos amounts to destroying human life. In reporting on this decision, Trevor Hughes and Minnah Arshad (2024) from *USA Today* recount the following argument made by Christian religious conservatives:

> On one side of the battle are millions of Americans who used in-vitro fertilization to have desperately wanted children who could otherwise not have been born. On the other are Christian religious conservatives who argue *humans should not be playing God in a laboratory*, and that life begins at conception (emphasis added).

This is a "playing God" argument against in vitro fertilization (IVF) as a form of Assisted Reproductive Technologies (ART). As Felix De Bie et al. (2023, 72) observe, "IVF is perceived by some as interfering with nature or *playing God*" (emphasis added).

Accordingly, those who claim that "humans should not be playing God in a laboratory" seem to be advancing a "playing God" argument against having children through IVF. Using my argumentation scheme for "playing God" arguments, such a "playing God" argument against IVF can be stated as follows.

Major Premise: We should not (morally speaking) "play God."
Minor Premise: Using IVF to have children amounts to "playing God."
Conclusion: We should not (morally speaking) use IVF to have children.

Before the conclusion of this argument can be accepted or rejected, we should evaluate this argument by asking the Critical Questions that accompany the argumentation scheme for "playing God" arguments. These questions are as follows.

CQ1: Could IVF get out of the control of (less than omnipotent) human beings?
CQ2: Could IVF be misunderstood or have unforeseen consequences for (less than omniscient) human beings?
CQ3: Could IVF be misused or abused by (less than omnibenevolent) human beings?

If the answer to all of these Critical Questions is "yes," then it means that using IVF to have children would indeed amount to "playing God" in terms of power, knowledge, and benevolence. Given that we should not pretend to be all-powerful, all-knowing, and all-good as God is supposed to be, we could then conclude that we should not use IVF to have children. In that case, the "playing God" argument against IVF would be a strong argument. On the other hand, if the answer to all of these Critical Questions is "no," then it means that using IVF to have children would not amount to "playing God." In that case, the conclusion that we should not use IVF to have children would not be warranted and the "playing God" argument against IVF would be a weak argument. Any other combination of "yes" and "no" answers to the Critical Questions, CQ1–CQ3, would make the "playing God" argument against IVF stronger or weaker. That is, the more Critical Questions are answered in the affirmative, the stronger the argument, whereas the more Critical Questions are answered in the negative, the weaker the argument (see Table 3.1 in Chapter 3).

First, let's consider CQ1. Could IVF get out of the control of (less than omnipotent) human beings? IVF could give infertile people the power to have children. As a form of ART, IVF involves fertilizing an egg with sperm in a laboratory dish, instead of inside the body, to create an embryo. The sperm and eggs can come from the parents-to-be or from donors. The embryo is then implanted in a uterus for pregnancy to take place. The embryo can be implanted in the uterus of the woman from whom the eggs were retrieved or that of another woman (a so-called "gestational surrogate" or "gestational carrier").[1] IVF has been around since 1978 when the first so-called "test tube baby," namely, Louise Brown, was born.[2] Since then, IVF has been used as a treatment for infertility. While IVF is only one form of ART, it is informative to note that approximately 2.3 percent of all infants born in the United States every year are conceived using ART (Centers for Disease Control and Prevention 2023). According to the CDC's Fertility Clinic Success Rates Report, "approximately 238,126 patients had 413,776 ART cycles performed at 453 reporting clinics in the United States, resulting in 112,088 clinical pregnancies, 91,906 live births (deliveries of one or more living infants), and 97,128 live born infants" (Centers for Disease Control and Prevention 2023, 12).

There is a sense in which IVF can be an empowering technology only for some but not for others. At least in the United States, fertility treatments can be very expensive. According to the US Department of Health and Human Services (2024), "costs for a single cycle of IVF have recently been estimated to range from $15,000 to $20,000 and can exceed $30,000. Given the average number of cycles that are needed to become pregnant from IVF is 2.5, this means that the average cost of IVF to conceive successfully can easily exceed $40,000." What's more, American health insurance companies do not always cover fertility treatments. According to the US Department of Health and Human Services (2024):

> IVF treatment is often not covered by insurance, so *the cost of treatment can pose a significant access challenge for lower income individuals.* The majority of states (29 states) do not require private insurance to cover any fertility services. Select services are required in 21 states and 15 states include IVF coverage mandates. Further, these state mandates do not apply to self-insured plans, which include 61% of workers. Medicaid coverage of fertility diagnosis or treatment is optional for states and coverage varies from state to state. Some states, like New York and Utah, choose to cover diagnosis and treatment of fertility services through Medicaid, while other states only cover diagnosis (emphasis added).

For these reasons, IVF can be unaffordable for many low-income Americans. Still, it could be argued that the unaffordability of IVF is the result of the time and place in which the technology is situated rather than an *intrinsic* feature of the technology itself. In other countries, IVF is subsidized by governments. For example, in Israel, "the world capital of in vitro fertilization," the Israeli government "provides free, unlimited IVF procedures for up to two 'take-home babies' until a woman is 45" (Kraft 2011). Even in the United States, some states have IVF coverage mandates, which in principle could become federal law. So, if more US states or American health insurance companies decided to cover IVF, then more people who need IVF would be able to access it. On the other hand, a reverse trend could also occur whereby IVF would become less affordable in more US states and perhaps even in other countries around the world. In that sense, the unaffordability of IVF could get out of control, which in turn would make IVF an empowering technology for the wealthy but not for others.

Nevertheless, one intrinsic aspect of IVF that may be considered as getting out of control is the growing number of frozen embryos stored in fertility clinics. Unlike affordability, this is an intrinsic aspect of IVF because, as we have seen, "the average number of cycles that are needed to become pregnant from IVF is 2.5" (US Department of Health and Human Services 2024), which means that IVF often results in leftover embryos that are then cryopreserved and sometimes remain frozen for many years. In other words, unless the success rate of IVF somehow becomes optimal, such that a single embryo transfer would be sufficient to achieve a live birth, creating leftover embryos will continue to be an intrinsic part of the IVF process, and the number of cryopreserved human embryos will continue to grow. According to the US Department of Health and Human Services (2024), "A large proportion of all ART cycles involve freezing egg or embryos. In 2021, nearly 41% of ART cycles (167,689 cycles) were egg or embryo banking cycles in which all resulting eggs or embryos were frozen for future use." This may be considered a morally unacceptable out-of-control result of IVF not only by those who believe that human embryos are children, as the Alabama Supreme Court has ruled, but also by those who believe that human embryos have some moral worth, even if not a moral worth that is equivalent to the moral worth of human beings after birth.

From all of the above, it is reasonable to conclude that IVF could get out of the control of (less than omnipotent) human beings. As long as IVF requires creating leftover human embryos that need to be cryopreserved, the number of frozen human embryos will only continue to increase and, in that sense, get out of control. Accordingly, the answer to CQ1, as far as IVF is concerned, seems to be "yes" overall.

Next, let's consider CQ2. Could IVF be misunderstood or have unforeseen consequences for (less than omniscient) human beings? IVF is considered a fairly safe and effective treatment for infertility nowadays. In 1969, Robert Edwards succeeded in fertilizing a human egg in a test tube, but failed to get the fertilized egg to develop beyond a single cell division. With the help of Patrick Steptoe, who used a laparoscope to remove eggs from ovaries, Edwards was able to put the eggs in cell culture, add sperm, and get the fertilized eggs to divide several times until embryos were formed. Following the birth of Louise Brown in 1978, IVF has been refined and improved numerous times, such that it is now an established therapy for infertility.

Like other medical interventions, however, IVF may have some complications for women. Some women who take fertility medication to stimulate egg development may experience allergic reactions. There is also the risk of Ovarian Hyperstimulation Syndrome (OHSS), a condition in which the ovaries swell and leak fluid into the abdominal cavity, which can be life-threatening on rare occasions. The process of egg retrieval poses some risk of infection and twisting of the ovary as well. Pregnancy complications include multiple pregnancies, especially if more than one embryo is transferred, ectopic pregnancy, heterotopic pregnancy, and placental abruption. Although some of these complications, such as ectopic pregnancy, may occur even without IVF. Overall, such potential complications are rare (US Department of Health and Human Services 2024).

While the potential complications from IVF for women are fairly well understood, there remains an open question among medical researchers "with regards to the outcome for the children" born through IVF (Supramaniam and Mittal 2018). More specifically, neurodevelopmental outcomes for children born through IVF as compared to children born without ART are still not very well understood. As Amaia Cavero-Ibiricu et al. (2024, 116) put it, "Despite the significant advances that have occurred in the last 40 decades in this field, the effect of ART on the developing human brain is still unclear, as well as the relationship they may have in the development of cerebral palsy (CP)." Some studies suggest "that the neurodevelopment of children born after fertility treatment is overall comparable to that of children born after spontaneous conception" (Bay, Mortensen, and Kesmodel 2013, 850), whereas other studies suggest that "children conceived with ART have a slightly increased risk of developing CP [i.e., Cerebral Palsy] compared with SC [i.e., spontaneously conceived] children" (Cavero-Ibiricu et al. 2024, 122).

From all of the above, it is reasonable to conclude that IVF could be misunderstood or have unforeseen consequences for (less than omniscient) human beings. As long as the long-term effects for children born through IVF are not well understood, and "further research of high methodological quality in children beyond pre-school age is needed" (Middleburg et al. 2008, 219), we cannot conclude with much confidence that we know everything there is to know about IVF and its long-term effects. Accordingly, the answer to CQ2, as far as IVF is concerned, seems to be "yes" overall.

Let's move on to CQ3. Could IVF be misused or abused by (less than omnibenevolent) human beings? There are several ways in which IVF can be misused or abused. For instance, IVF could be abused through commercial (or paid) surrogacy. That is, a woman could be paid to carry to term an embryo that was created through IVF from donor eggs and sperm. From a clinical perspective, this type of surrogacy is referred to as "gestational surrogacy" as the surrogate has no genetic relation to the fetus that is undergoing gestation in her womb. To some, paid surrogacy is morally objectionable because it "can result in the commodification of human life and the exploitation of women" (Piersanti et al. 2021, 2). In fact, so-called "reproductive tourism," "procreative tourism," or "fertility tourism" is already a well-known phenomenon whereby individuals and couples "travel across national and international borders with the intention of receiving medical advice, assisted reproductive technology (ART) treatments, and, in some cases, donor gametes, embryos, or surrogacy services" (Inhorn and Patrizio 2009, 904).

For instance, until 2021, India was among a handful of countries, including Georgia, Russia, Thailand, and Ukraine, where commercial (or paid) surrogacy is permitted. (In the United States, there are no federal laws governing surrogacy, so state laws, if any, vary greatly from legally sanctioned surrogacy contracts in some states to outright bans in other states.) A Reuters article on India's "booming 'rent-a-womb' industry" from 2013 reads as follows (Bhalla and Thapliyal 2013):

> India's surrogacy industry is vilified by women's rights groups who say *fertility clinics are nothing more than "baby factories" for the rich.* In the absence of regulation, they say many *poor and uneducated women are lured by agents,* hired by clinics, *into signing contracts they do not fully understand* (emphasis added).

The use of the phrases "rent-a-womb industry" and "baby factories" suggests that paid surrogacy is deemed morally objectionable when "poor and uneducated women" are exploited by fertility clinics and surrogacy agencies to provide rich people with the babies they want so desperately. The technology that makes paid surrogacy possible in the first place, namely, IVF, can thus be abused when fertility clinics and surrogacy agencies operate as "baby factories." For example, one of the world's most popular surrogacy agencies, BioTexCom, which is based in Kyiv, continues to operate at full capacity despite allegations of mishandling embryos, faking DNA tests, child trafficking, and baby selling (Gridneff, Schultheis, and Drabyk 2023). Even the war with Russia has not slowed down the operation of this Ukrainian surrogacy agency. With an ever-growing demand, the "number of providers [creating embryos for sale] remains wholly unknown" (Klitzman and Sauer 2015, 167).

IVF could be misused or abused by not only medical providers but also parents. Parents with a terminally ill child may choose to have another child through IVF that will serve as a so-called "savior sibling." The term refers "to children intentionally created to donate biological material, most commonly cord blood, but theoretically also bone marrow or solid organs (liver and kidney), to save the life of an already-existing child" (Althorpe and Finneron-Burns 2023, 49–50). In this case, IVF is used to create multiple embryos, and then preimplantation genetic diagnosis (PGD) is used to select the ones that are a human leukocyte antigen (HLA) match for the already-existing, sick child. The creation of savior siblings is fraught with many ethical challenges, which is why some conclude that the practice is morally reprehensible. Chief among these moral concerns are that creating a savior sibling is (a) the wrong reason to bring a child into the world, (b) exploitative insofar as the created child is being used as a mere means (i.e., providing a cure) to an end (i.e., curing an existing sick child), (c) potentially harmful to the created child, and (d) is devaluing of human dignity and human life (Shapiro 2018).

IVF could lead to further abuse insofar as it is the technology that enables germline gene editing. Recall the "playing God" argument against genetic engineering discussed as an example in Chapter 2. When gene editing is done at the level of germline cells (i.e., eggs and sperm cells), it could introduce unwanted modifications to the human genome that would be inheritable. Using IVF for germline gene editing would constitute an abuse of IVF because the technology was not intended to enable such genetic engineering of human germline cells. Indeed, in 2018, there was moral outrage in the biomedical community when a Chinese biomedical researcher, He Jiankui, announced his "use of the genome editor CRISPR to alter the DNA of human embryos and implant them into two women led to three births, sparking ethical outrage and fears for the babies' health (about which little is known)" (Cohen 2022). Accordingly, IVF could be abused by enabling the irresponsible use of gene editing tools, such as CRISPR, for the purpose of embryo editing.

From all of the above, it is reasonable to conclude that IVF could be misused or abused by (less than omnibenevolent) human beings. In the hands of unscrupulous fertility clinics and surrogate agencies, embryos and even babies can be created through IVF just for the sake of making money at the expense of women in dire financial straits and people who desperately want to have children of their own. Some parents may even choose to have a child through IVF simply for that child to serve as a "savior sibling" for an existing, sick child. IVF could lead to further abuse by making germline gene editing feasible. Accordingly, the answer to CQ3, as far as IVF is concerned, seems to be "yes" overall.

As far as IVF is concerned, Critical Questions CQ1–CQ3 seem to have affirmative answers. This means that using IVF to have children would indeed amount to "playing God" in terms of power (CQ1), knowledge (CQ2), and benevolence (CQ3). Therefore, a "playing God" argument against IVF is a strong argument (see Table 3.1 in Chapter 3). This argument can be summed up as follows.

Major Premise: We should not (morally speaking) "play God."
Conditional Premise: If IVF could get out of the control of humans, be misunderstood or have unforeseen consequences for humans, or be misused or abused by humans, then using IVF to have children amounts to "playing God."
Control Premise: IVF could get out of the control of humans.
Understanding Premise: IVF could be misunderstood or have unforeseen consequences for humans.
Use Premise: IVF could be misused or abused by humans.
Conclusion: We should not (morally speaking) use IVF to have children.

This strong "playing God" argument against IVF points to a few moral problems with IVF as a form of ART, namely, that human embryo biobanking could get out of control, that IVF could have unintended long-term consequences for those born through IVF, which we do not understand yet, and that IVF could be misused or abused by medical providers and parents.

Like all "playing God" arguments, this one against IVF is a defeasible argument. As we have seen in Chapter 3, conclusions of defeasible arguments can always be defeated by additional evidence that is not cited in their premises. In other words, this argument is not the final word about whether or how we should use IVF as a form of ART. One obvious way to defeat this "playing God" argument against IVF is to argue that the potential benefits of IVF would outweigh the perils of loss of control, misunderstanding, unforeseen consequences, misuse, and abuse. Still, this strong "playing God" argument against IVF gives us pretty good reasons to be concerned about this technology, at least until we can make sure that we are able to control it, understand it and its effects, and use it benevolently.

Another new or emerging technology that broadly falls under the category of Assisted Reproductive Technologies (ART), insofar as it might someday become part of Neonatal Intensive Care (NIC), is the so-called "artificial womb." According to De Bie et al. (2023), "the EXTra-uterine Environment of Neonatal Development (EXTEND) artificial womb technology (AWT) system" (67) provides "a womb-like incubation, ensuring oxygenation via an oxygenator connected to the fetal umbilical vessels in addition to providing nutrition and a warm, insulated, fluid-filled environment supporting fetal hemodynamic physiology and continued maturation" (68). Concerns about "playing God" have already been expressed with respect to AWT. For instance, an article in the magazine, *Relevant*, titled "An 'Artificial Womb Facility' That Gives Parents the Ability to Pick Their Child's Traits Is in the Works," starts as follows:

> Well, people are *playing God* again. The world's first "artificial womb facility" is in the works, and it plans to give clients the ability to choose their baby's characteristics from a menu. EctoLife was conceptualized by Hashem Al-Ghaili, a Berlin-based biotechnologist, and he believes his plan could be beneficial for couples struggling with conception for any reason. EctoLife would work by *taking embryos and placing them* in "growth pods," or *artificial wombs*, which would be *designed to replicate the exact conditions of a mother's uterus* (emphasis added).

An investigation by the Associated Press revealed that the alleged "artificial womb facility" is not real (Tulp 2022). In an email to the Associated Press, Al-Ghaili himself confirmed that EctoLife is not a real company. In his email to the Associated Press, Al-Ghaili wrote, "EctoLife is NOT a real facility and there is no work being done to create the prototype as of now.[3] The video is intended to showcase how far science and reproductive technology have progressed and initiate the discussion around such technology" (Tulp 2022).[4]

So, let's have a discussion about AWT. After all, Dr. Christoph Haller, a pediatric heart surgeon at The Hospital for Sick Children in Toronto, and his team have already tested their AWT in animal models, specifically, pigs, and "their hope is that someday, [AWT] will help humans survive extremely premature birth and avoid serious complications, such as blindness and permanent damage to lungs and brains" (Stein 2024). Conversely, there are those who find AWT morally objectionable "because it seems like it's playing God, that machines could be preferable to a woman's womb" (Hand to Hold 2017), which looks like a "playing God" argument against AWT. As another example of a biomedical technology, in addition to IVF, it would be useful to see how my argumentation scheme for "playing God" arguments and its accompanying Critical Questions can handle such an example.

Using my argumentation scheme for "playing God" arguments, then, such a "playing God" argument against AWT can be stated as follows.

Major Premise: We should not (morally speaking) "play God."
Minor Premise: Using AWT to incubate human fetonates amounts to "playing God."
Conclusion: We should not (morally speaking) use AWT to incubate human fetonates.[5]

Before the conclusion of this argument can be accepted or rejected, we should evaluate this argument by asking the Critical Questions that accompany my argumentation scheme for "playing God" arguments. These questions are as follows.

CQ1: Could AWT get out of the control of (less than omnipotent) human beings?
CQ2: Could AWT be misunderstood or have unforeseen consequences for (less than omniscient) human beings?
CQ3: Could AWT be misused or abused by (less than omnibenevolent) human beings?

If the answer to all of these Critical Questions is "yes," then it means that using AWT to incubate human fetonates would indeed amount to "playing God" in terms of power, knowledge, and benevolence. Given that we should not pretend to be all-powerful, all-knowing, and all-good as God is supposed to be, we could then conclude that we should not use AWT to incubate human fetonates. In that case, the "playing God" argument against AWT would be a strong argument. On the other hand, if the answer to all of these Critical Questions is "no," then it means that using AWT to incubate human fetonates would not amount to "playing God." In that case, the conclusion that we should not use AWT to incubate human fetonates would not be warranted and the "playing God" argument against AWT would be a weak argument. Any other combination of "yes" and "no" answers to the Critical Questions, CQ1–CQ3, would make the "playing God" argument against AWT stronger or weaker. That is, the more Critical Questions are answered in the affirmative, the stronger the argument, whereas the more Critical Questions are answered in the negative, the weaker the argument (see Table 3.1 in Chapter 3).

First, let's consider CQ1. Could AWT get out of the control of (less than omnipotent) human beings? Much like IVF, given the increasing demand for

ART, market forces could make the cost of AWT such that it would be out of reach for low-income people. However, as in the case of IVF, it could be argued that, if AWT were to become unaffordable, and thus inaccessible to low-income people, it would be the result of the time and place in which the technology would be situated rather than an *intrinsic* feature of the technology itself.

Nevertheless, there is a sense in which AWT could rob people who want to have children, but need to resort to ART, of their decision-making power and autonomy. Consider, if you will, a woman who gives birth to a fetal neonate. The medical staff then instructs her that her fetonate must be put in an artificial womb. Suppose further that this woman is wary of AWT and its potential harmful effects on her newborn. Would she be able to refuse incubating her newborn in an artificial womb? Or would the pressure to consent be rather difficult, perhaps even impossible, to resist? As Vardit Ravitsky and Louise King (2024) put it:

> In a political climate where reproductive rights are being curtailed in the name of protecting fetuses, safe and effective artificial wombs could be used to argue that the fetus not only has a right to live, but also has the right to any and all medical care available. Such an argument could promote legislation that requires *dangerous surgery* [e.g., cesarean sections] *despite a patient's objection*. With current legislation that defines personhood in various states, this is a real possibility (emphasis added).

In that sense, AWT could be used in ways that limit reproductive freedom.[6] But again, it could be argued that the use of AWT to limit reproductive rights would be the result of the time and place in which the technology would be situated rather than an *intrinsic* feature of the technology itself. After all, given a different political climate, the reproductive rights of women could be respected rather than curtailed in order to optimize outcomes for the newborn at the expense of the mother. If AWT were to be considered medical research, or "experimental AWT," rather than medical treatment, as Elizabeth Chloe Romanis (2020, 400) argues, then the technology would only be tested under the rigorous conditions of clinical research, which include "randomization, blinding, and controls."

Along similar lines, one could also imagine a sort of Brave New World in which AWT replaces natural pregnancies altogether. In this imaginary world of ectogenesis, all human fetuses are gestated in artificial wombs rather than human bodies. But again, if AWT were to become a mandatory replacement for human wombs, it would be the result of the time and place in which the technology would be situated, or the political climate, rather than an *intrinsic* feature of the technology itself. Indeed, in Aldous Huxley's dystopian novel, ectogenesis is the culmination of a political vision in which the government (World State) exercises control over every aspect of human life, including human reproduction.[7]

From all of the above, it is reasonable to conclude that AWT, that is, the technology itself, could not get out of the control of (less than omnipotent) human beings. However, the *affordability* of AWT could get out of control, which would make the technology the privilege of a select few and put it beyond the reach of low-income people. Moreover, depending on the political climate, AWT could be used to uphold the rights of newborns over those of mothers by requiring medical procedures that benefit the fetonate even when the mother does not consent. Rather than AWT getting out of human control altogether, however, this would be an instance of some people using AWT to exercise control over others in order to bring their preferred political vision to fruition. Accordingly, the answer to CQ1, as far as AWT is concerned, seems to be "no" overall.

Next, let's consider CQ2. Could AWT be misunderstood or have unforeseen consequences for (less than omniscient) human beings? As we have seen in Chapter 3, new or emerging technologies tend to be disruptive. Disruptive technologies, by their very nature, could be misunderstood by, and have unforeseen consequences for, less than all-knowing beings like us. Now, AWT seems to fit the description of a disruptive technology. According to John Danaher and Jeroen Hopster (2022, 10), "Various other potentially disruptive technologies, such as autonomous weapon systems, solar radiation management, or the artificial womb, may similarly give rise to harmful uncertainty in defined domains of inquiry." Therefore, by virtue of being a disruptive technology, AWT could be misunderstood by, or have unforeseen consequences for, human beings.

One of the main points of confusion surrounding AWT has to do with the moral and legal status of the subject treated with AWT. Would the subject treated with AWT be considered a fetus? Would it be considered a newborn baby? Should it be considered a patient? If so, should it have the same rights as other patients? As Mark Mercurio, Director of the Program for Biomedical Ethics at the Yale School of Medicine, explained in an interview with NPR, whether we call the subjects treated with AWT "fetuses," "newborns," or "babies" matters "in terms of how we assign moral status to individuals. How much their interests—how much their welfare—should count. And what one can and cannot do for them or to them" (Stein 2024). According to De Bie et al. (2023, 74), "the best way to describe the person who would receive current AWT is as a 'fetal neonate' or fetonate. Neonatal pertains to the fact that the subject is removed from the womb, hence is newly (*neo*) born (*natus*). At the same time, the core objective of the AWT is to conserve fetal physiology, justifying the use of the modifying term fetal." Romanis (2023, 90) disagrees because she thinks that combining the terms "fetus" and "neonate" is oxymoronic. Currently, we do not have answers to these fundamental conceptual questions about the subject of AWT.[8]

As in the case of IVF, there is much that we, who are not omniscient, still do not understand about the long-term effects of AWT. For, as De Bie et al. (2023, 72) note, "long-term effects are *hard to investigate* in experimental animal models, hence will *remain unanswered* until well after the first human research subjects reach an age allowing for these assessments" (emphasis added). For instance, like other medical interventions, AWT could have various unforeseen complications. According to De Bie et al. (2023, 72), for fetonates, the

> primary harm of AWT at this stage would be the risk of death or severe disability caused by AWT or its complications. Severe negative effects secondary to AWT could include *short- and long-term physiological effects*, e.g., intracranial hemorrhage due to heparinization of the fetonate on AWT. In addition, *potential psychological and behavioral effects* secondary to the lack of physical maternal-fetal bonding *should be thoroughly examined* (emphasis added).

Notice that these potential complications, risks, and negative effects "should be thoroughly examined," which means that they are not well understood yet.

As far as complications for the mother are concerned, if AWT were to require fetal extraction by means of C-sections, it could lead to "a higher perioperative risk for maternal complications (such as bleeding, complicated extraction, higher risk of uterine rupture in future pregnancies) at earlier stages of pregnancy" (De Bie et al. 2023, 73). Moreover, "the uncertainty of AWT may exacerbate the emotional and psychological burden of unpredictable fetal morbidity on the parents" (De Bie et al. 2023, 73). More generally, it is difficult to tell how AWT would change the experience of pregnancy for mothers. For instance, the "lack of physical maternal-fetal bonding during the fetonate's course of treatment with AWT" (De Bie et al. 2023, 73) could detract from the experience of motherhood as a meaningful and self-fulfilling personal experience.

From all of the above, it is reasonable to conclude that AWT could be misunderstood or have unforeseen consequences for (less than omniscient) human beings. Currently, we do not understand how AWT could affect both the mother and the fetonate, which means that AWT could have many complications, risks, and long-term effects that would be unforeseen to us, who are not all-knowing as God is supposed to be. We also do not have a clear grasp of the moral and legal status that subjects treated with AWT would or should have. Accordingly, the answer to CQ2, as far as AWT is concerned, seems to be "yes" overall.

Let's move on to CQ3. Could AWT be misused or abused by (less than omnibenevolent) human beings? As we have seen in Chapter 3, some technologies are dual use technologies insofar as they can be (mis)used (or abused) in ways other than their intended or designed use. AWT is a dual use technology insofar as it can be used for medical purposes, for example, as treatment for prematurity or pathological pregnancies, which are its intended purposes, and for non-medical purposes, which are not its intended purposes. The latter would be a misuse of AWT in cases in which the technology is used to avoid pregnancy altogether. More specifically, people who want to have children but do not want to be pregnant could use AWT to avoid pregnancy and have a fetus gestate in an artificial womb instead of a uterus (De Bie et al. 2023, 72).

In that respect, even though it "is not the intended endpoint of current research, [AWT] could eventually lead to 'full ectogenesis'—growing a foetus from conception to 'birth' wholly outside the human body" (Wilkinson, Williams, and Fovargue 2023). As De Bie et al. (2023, 75) observe, so-called "full ectogenesis" or "complete ectogenesis," whereby the entire process from conception to birth would occur outside of a human body, "could lead to the *devaluation or even pathologizing of pregnancy*, and may *diminish women's experience of deriving meaning, empowerment, and self-fulfillment* from this unique aspect of female biology" (emphasis added). Moreover, "AWT could become a tool of coercion guided by the idea that women regarded as 'substandard gestators' could be pressured to use AWT for the safety of the fetus, hence violating maternal autonomy" (De Bie et al. 2023, 75).

The widespread use of AWT, as depicted in the fictional video about EctoLife,[9] could also result in the devaluation of human life insofar as the (mis)use of AWT for full or complete ectogenesis would be dehumanizing. The moral concern here is similar to the one typically expressed with regards to human cloning. As Leon Kass (1997) argues, human cloning replaces natural procreation with the artificial manufacture of children as products, which turns it into a demeaning process that erodes our respect for human beings. For this reason, Kass (1997, 23) argues, human cloning "is profoundly dehumanizing." Similarly, it could be argued that the use of AWT for complete ectogenesis would replace natural procreation with the artificial manufacture of children as products, thereby turning it into a demeaning process that would erode our respect for human life. In that respect, the use of AWT for complete ectogenesis would count as a form of abuse insofar as it would be dehumanizing. Along similar lines, Rosemarie Tong claims that AWT "could lead to a commodification of the whole process of pregnancy" (quoted in Rosen 2003).

From all of the above, it is reasonable to conclude that AWT could be misused or abused by (less than omnibenevolent) human beings. AWT could be (mis)used for non-medical reasons, such as avoiding pregnancy, instead of medical reasons, such as treating prematurity or pathological pregnancies. AWT could be abused insofar as full or complete ectogenesis would be dehumanizing, leading to the commodification of pregnancy, and thus result in the devaluation of human life. Accordingly, the answer to CQ3, as far as AWT is concerned, seems to be "yes" overall.

As far as AWT is concerned, Critical Question CQ1 seems to have a negative answer, whereas Critical Questions CQ2 and CQ3 seem to have affirmative answers. This means that using AWT to incubate human fetonates would indeed amount to "playing God" in terms of knowledge (CQ2) and benevolence (CQ3), but not in terms of power (CQ1). Therefore, a "playing God" argument against AWT is a somewhat strong argument (see Table 3.1 in Chapter 3). This argument can be summed up as follows.

Major Premise: We should not (morally speaking) "play God."
Conditional Premise: If AWT could get out of the control of humans, be misunderstood or have unforeseen consequences for humans, or be misused or abused by humans, then using AWT to incubate human fetonates amounts to "playing God."
Control Premise: AWT could not get out of the control of humans.
Understanding Premise: AWT could be misunderstood or have unforeseen consequences for humans.
Use Premise: AWT could be misused or abused by humans.
Conclusion: We should not (morally speaking) use AWT to incubate human fetonates.

While this "playing God" argument against AWT is only somewhat strong, it still points to a few moral problems with AWT, namely, that it could be misunderstood, have unforeseen complications and long-term effects, and be misused or abused by both parents and medical providers.

Like all "playing God" arguments, this one against AWT is a defeasible argument. As we have seen in Chapter 3, conclusions of defeasible arguments can always be defeated by additional evidence that is not cited in their premises. In other words, this argument is not the final word about whether or how we should use AWT. One obvious way to defeat this "playing God" argument against AWT is to argue that the potential benefits of AWT would outweigh the perils of misunderstanding, unforeseen consequences, misuse, and abuse. Still, this somewhat strong "playing God" argument against AWT gives us pretty good reasons to be concerned about this technology, at least until we can make sure that we are able to understand it and its effects as well as use it benevolently.

As we have seen in Chapter 3, our answers to the Critical Questions, CQ1–CQ3, will tell us what issues need to be addressed before the new or emerging technology in question can be developed, deployed, or used responsibly, that is, in accordance with our human limitations and without pretending that we are all-powerful, all-knowing, and all-good. Recall the example of a "playing God" argument against using genetic engineering to bring back extinct species or genetically modify extant species from Chapter 3.

Major Premise: We should not (morally speaking) "play God."
Minor Premise: Using genetic engineering to bring back extinct species or genetically modify extant species amounts to "playing God."
Conclusion: We should not (morally speaking) use genetic engineering to bring back extinct species or genetically modify extant species.

As far as this "playing God" argument is concerned, the Critical Questions, CQ1–CQ3, are as follows.

CQ1: Could the genetic engineering of animals get out of the control of (less than omnipotent) human beings?
CQ2: Could the genetic engineering of animals be misunderstood or have unforeseen consequences for (less than omniscient) human beings?
CQ3: Could the genetic engineering of animals be misused or abused by (less than omnibenevolent) human beings?

Suppose that our answer to CQ1 and CQ2 is "no," but our answer to CQ3 is "yes." In that case, we would have a somewhat weak "playing God" argument against using genetic engineering to bring back extinct species or genetically modify extant species. Nevertheless, we would have reason to believe that, as far as genetic engineering technology is concerned, the main moral concern is not that it could get out of control, be misunderstood, or have unintended, negative consequences. Rather, the main moral concern is that it could be misused or abused by human beings who, unlike a supposedly omnibenevolent being, are not all-good. Having such a reason would help us to direct our governance efforts toward crafting policies, regulations, or laws designed to prevent misuse or abuse instead of more research (to address CQ2) and safety testing (to address CQ1).

In general, an affirmative answer to CQ1 suggests that the new or emerging technology in question could get out of the control of (less than omnipotent) human beings. Of course, we could never have perfect control over a new or emerging technology because we are not all-powerful as God is supposed to be. However, we can try to enhance our control over a new or emerging technology by analyzing risks and designing for safety. Accordingly, a "yes" to CQ1 suggests that we should direct our efforts toward risk analysis and safety. That is, we should conduct extensive risk analyses and safety testing, as well as design safety mechanisms informed by the results of these risk assessments and safety tests, to make sure that the new or emerging technology in question would be controllable by us before it can be deployed or used responsibly, that is, in accordance with our human limitations and without pretending that we are all-powerful.

An affirmative answer to CQ2 suggests that the new or emerging technology in question could be misunderstood or have unforeseen consequences for (less than omniscient) human beings. Of course, we could never have perfect understanding of a new or emerging technology and know all of its potential impacts because we are not all-knowing as God is supposed to be. However, we can try to improve our knowledge and understanding of a new or emerging technology through further research. Accordingly, a "yes" to CQ2 suggests that we should direct our efforts toward research. That is, we should conduct extensive experiments and simulations to learn as much as we can about the new or emerging technology in question before it can be deployed or used responsibly, that is, in accordance with our human limitations and without pretending that we are all-knowing.

An affirmative answer to CQ3 suggests that the new or emerging technology in question could be misused or abused by (less than omnibenevolent) human beings. Of course, we could never deploy or use a new or emerging technology in perfectly good ways because we are not all-good as God is supposed to be. However, we can try to prevent misuse and abuse by enacting regulations, policies, and laws with robust enforcement mechanisms. Accordingly, a "yes" to CQ3 suggests that we should direct our efforts toward policy. That is, we should conduct extensive deliberations, which would include all stakeholders, aimed at drafting regulations or legislating laws pertaining to the new or emerging technology in question before it can be deployed or used responsibly, that is, in accordance with our human limitations and without pretending that we are all-good.

To sum up, here is how our answers to Critical Questions, CQ1–CQ3, can provide some general guidance for approaching the governance of a new or emerging technology.

CQ1: Could *T* get out of the control of (less than omnipotent) human beings?

- If the answer is "no," move on to CQ2.
- If the answer is "yes," attempt to enhance control by focusing on safety through risk assessment, safety testing, etc.

CQ2: Could *T* be misunderstood or have unforeseen consequences for (less than omniscient) human beings?

- If the answer is "no," move on to CQ3.
- If the answer is "yes," attempt to increase knowledge and understanding of consequences by focusing on research through experimentation, simulation, etc.

CQ3: Could *T* be misused or abused by (less than omnibenevolent) human beings?

- If the answer is "no," stop and review your answers to CQ1–CQ3.
- If the answer is "yes," attempt to prevent misuse or abuse by focusing on policy through regulation, legislation, enforcement, etc.

Now, as far as the "playing God" argument against AWT is concerned, we have answered Critical Questions, CQ2 and CQ3, in the affirmative, but not CQ1. This gives us good reasons to be concerned about AWT being misunderstood or having unintended, negative consequences, and being misused or abused, but not getting out of control. Having good reasons to believe this would, in turn, help us to direct our governance efforts as far as AWT is concerned. First, our affirmative answer to CQ2 suggests that we should direct our governance efforts toward further research on our notions of motherhood, parenthood, and the like, as well as the potentially harmful impacts of AWT on both mothers and fetonates. Second, our affirmative answer to CQ3 suggests that we should direct our governance efforts toward drafting policies or legislating laws designed to prevent the misuse or abuse of AWT by parents and medical providers. In that respect, the "playing God" argument against AWT, along with our answers to the Critical Questions, CQ1–CQ3, may provide some general guidance to AWT governance.[10]

When the phrase "playing God" is used in debates concerning the ethical development, deployment, or use of new or emerging technologies, it is usually interpreted as a warning not to interfere with God's creation or nature. As we have seen in Chapter 2, this interpretation of "playing God" arguments as a call to non-interference with nature may be too narrow, which means that an alternative interpretation of "playing God" arguments is required. In Mizrahi (2020), I articulate an interpretation of "playing God" arguments according to which such arguments invite us to think carefully about all the ways in which the development, deployment, or use of new or emerging technologies could go seriously morally wrong, given the fact that we are not as powerful, wise, and good as God is supposed to be. On my interpretation, then, "playing God" arguments are a critique of humanity rather than an appeal to anything supernatural like God. They are pleas to acknowledge human limitations, to avoid hubris and eventual humiliation, and to be cautious and humble instead.

Taking an argumentation theory approach, I propose the following argumentation scheme along with its accompanying Critical Questions. Together, they are meant to capture the moral concerns expressed by "playing God" arguments.

"Playing God" Argumentation Scheme

Major Premise: We should not (morally speaking) "play God."
Minor Premise: Developing, deploying, or using technology *T* amounts to "playing God."
Conclusion: We should not (morally speaking) develop, deploy, or use *T*.

Critical Questions

CQ1: Could *T* get out of the control of (less than omnipotent) human beings?

CQ2: Could *T* be misunderstood or have unforeseen consequences for (less than omniscient) human beings?
CQ3: Could *T* be misused or abused by (less than omnibenevolent) human beings?

This interpretation of "playing God" arguments as invitations to think carefully about all the ways in which developing, deploying, or using new or emerging technologies could go seriously morally wrong, rather than warnings to leave God's creation or nature alone, has at least four advantages. First, it provides a straightforward way to judge the strength of "playing God" arguments. Second, it allows us to make "playing God" arguments without thereby being committed to the existence of God or to any views about the creation of the universe. Third, it allows us to advance "playing God" arguments even when the new or emerging technology in question is not designed to intervene in nature directly. Fourth, it can provide some guidance as to where to direct our governance efforts. In that respect, my argumentation scheme for "playing God" arguments and its accompanying Critical Questions are meant to be conceptual tools for the analysis and evaluation of "playing God" arguments wherever they might occur. Taken together, the argumentation scheme for "playing God" arguments and its accompanying Critical Questions can provide a constructive lens through which to determine whether we need to make sure that we are able to control, understand, and benevolently use a new or emerging technology before it is morally permissible for us to deploy and use it.

In Part II of this book, I aim to demonstrate the usefulness of my argumentation scheme for "playing God" arguments and its accompanying Critical Questions as conceptual tools for the analysis and evaluation of "playing God" arguments. To this end, I analyze and evaluate four "playing God" arguments as use cases. The four use cases pertain to the following new or emerging technologies:

- Climate engineering or geoengineering (Chapter 5)
- Artificial Intelligence or AI (Chapter 6)
- Autonomous vehicles or self-driving cars (Chapter 7)
- Lethal Autonomous Weapon Systems or LAWS (Chapter 8)

For each of these new or emerging technologies, there are moral concerns or questions that are captured by "playing God" arguments. These questions are (see Chapter 1):

- Can we control it? (omnipotence)
- Do we understand it? (omniscience)
- Will we be corrupted by it? (omnibenevolence)

As we have seen in Chapter 1, these are fundamental questions about the limits of our power to control our environment, our knowledge of our world and our place in it, and our corruptible nature or moral character. As we go through these use cases, we should keep in mind that, while they can be strong arguments, "playing God" arguments are not meant to be conclusive, knockdown arguments. Rather, they are defeasible arguments. Accordingly, if the major and minor premises of a "playing God" argument are true, the conclusion of that argument would then be supported by the evidence cited in those premises, but the conclusion can always be defeated by additional evidence that is not cited in the premises.

Notes

1 I say more about surrogacy in relation to IVF below.
2 After Louise Brown was born, "Religious groups were opposed to the idea of 'playing God' with reproduction, and to a process in the course of which many embryos often died" (Nugent 2018).
3 The video can be watched here: https://youtu.be/O2RIvJ1U7RE?si =UyKkCgaMoXMMMuXr.
4 On the spread of misinformation on social media, see Chapter 3. On how AI can exacerbate the spread of misinformation on social media, see Chapter 6.
5 According to De Bie et al. (2023, 74), "the best way to describe the person who would receive current AWT is as a 'fetal neonate' or fetonate. Neonatal pertains to the fact that the subject is removed from the womb, hence is newly (*neo*) born (*natus*). At the same time, the core objective of the AWT is to conserve fetal physiology, justifying the use of the modifying term fetal." Cf. Elizabeth Chloe Romanis (2023).

6 On "playing God" and patient autonomy, see Vaughn (2023, 100–1).

7 I say more about full ectogenesis as a misuse of AWT, rather than AWT getting out of our control, below.

8 Relatedly, some of our notions of motherhood, parenthood, and the like may have to change as well. For example, as Wilkinson, Williams, and Fovargu (2023) point out, "In the UK, the person who gives birth is the child's legal mother—regardless of genetics or intention. Growing a foetus in an artificial womb could however sever this link between gestation and motherhood."

9 The video can be watched here: https://youtu.be/O2RIvJ1U7RE?si=UyKkCgaMoXMMMuXr.

10 It is important to note that the answers to the Critical Questions, CQ1–CQ3, of a "playing God" argument with respect to a new or emerging technology are meant to provide general guidance for approaching the governance of that technology. They are not supposed to constitute concrete governance recommendations or serve as a comprehensive governance proposal.

Part II

"Playing God" Arguments in Practice

5

Are We "Playing God" with Geoengineering?

The title of a recent article about climate engineering (or geoengineering) in *The Atlantic* reads, "Playing God with the Atmosphere" (Koren 2024). On the set of technologies commonly known as solar geoengineering in particular, Marina Koren (2024) writes, "Scientists *can't be sure* that *playing God* with the atmosphere won't cause human suffering, even if it is *intended to* alleviate it" (emphasis added). This is an example of a "playing God" argument not only because of the use of the phrase "playing God" but also because of the appeal to human limitations, specifically, epistemic (i.e., "can't be sure") and moral (i.e., "intended") ones. That is, even if we intend to do the morally right thing by attempting to mitigate the harmful effects of climate change by means of engineering the Earth's climate, for all we know, our good intentions might actually lead to more human suffering rather than less. After all, unlike God, a supposedly morally prefect being that can do no wrong, we are not morally prefect. Furthermore, even if we think that solar geoengineering can help to mitigate the effects of climate change, our understanding of these emerging technologies may be limited at best. After all, unlike God, a supposedly omniscient being that understands everything, we are not all-knowing.

Climate change is a serious problem that demands our immediate attention. Climate engineering (or geoengineering) is often proposed as a set of technological solutions, or so-called "techno-fixes," to climate change. These techno-fixes for counteracting the disastrous impacts of climate change include Solar Radiation Modification or Solar Radiation Management (SRM), which are technologies designed to cool down the Earth's atmosphere by means of injecting aerosol particles into the stratosphere to reflect sunlight (also known as stratospheric aerosol injection or SAI) or by means of seeding low-lying marine clouds with sea salt particles to enhance their brightness (also known as marine cloud brightening or MCB).[1] Techno-optimists would favor an uncritical adoption of solar geoengineering as a set of techno-fixes for climate change, whereas techno-pessimists would favor an uncritical rejection of solar geoengineering as a set of techno-fixes for climate change (see Chapter 1). For example, Mark Lynas (2011) claims that "playing God is good for the planet." On the other hand, Shannon Hall (2015) argues that "we can't keep playing God with the climate."

Instead of techno-optimism or techno-pessimism, I submit, what we need is a critical attitude with respect to solar geoengineering. Adopting a critical attitude means engaging with logical arguments. This is where my argumentation scheme for "playing God" arguments and its accompanying Critical Questions, CQ1–CQ3, may come in handy. They could prove helpful as a general framework for thinking critically about the morality of solar geoengineering. For, as we have seen, "playing God" arguments are already part of the debate about climate engineering (Koren 2024). As David Morrow (2020, 90) puts it, "Concerns about hubris and 'playing God' cut across all of these discussions [about solar geoengineering], as deliberately intervening in the climate provokes concerns about overstepping humanity's proper relationship with nature."

So, let's consider the "playing God" argument against solar geoengineering.[2] Using my argumentation scheme for "playing God" arguments (see Chapter 3), such a "playing God" argument against solar geoengineering can be stated as follows.

Major Premise: We should not (morally speaking) "play God."
Minor Premise: Developing and deploying solar geoengineering techno-fixes to counteract climate change amount to "playing God."
Conclusion: We should not (morally speaking) develop and deploy solar geoengineering techno-fixes to counteract climate change.

Before the conclusion of this argument can be accepted or rejected, we should evaluate this argument by asking the Critical Questions that accompany the argumentation scheme for "playing God" arguments. These questions are as follows.

CQ1: Could solar geoengineering get out of the control of (less than omnipotent) human beings?

CQ2: Could solar geoengineering be misunderstood or have unforeseen consequences for (less than omniscient) human beings?
CQ3: Could solar geoengineering be misused or abused by (less than omnibenevolent) human beings?

If the answer to all of these Critical Questions is "yes," then it means that developing and deploying solar geoengineering techno-fixes, such as SAI or MCB, to mitigate the negative impacts of climate change would indeed amount to "playing God" in terms of power, knowledge, and benevolence. Given that we should not pretend to be all-powerful, all-knowing, and all-good as God is supposed to be, we could then conclude that we should not develop and deploy solar geoengineering techno-fixes, such as SAI or MCB, to mitigate the negative impacts of climate change. In that case, the "playing God" argument against solar geoengineering would be a strong argument. On the other hand, if the answer to all of these Critical Questions is "no," then it means that developing and deploying solar geoengineering techno-fixes, such as SAI or MCB, to mitigate the negative impacts of climate change would not amount to "playing God." In that case, the conclusion that we should not develop and deploy solar geoengineering techno-fixes, such as SAI or MCB, to mitigate the negative impacts of climate change would not be warranted and the "playing God" argument against solar geoengineering would be a weak argument. Any other combination of "yes" and "no" answers to the Critical Questions, CQ1–CQ3, would make the "playing God" argument against solar geoengineering stronger or weaker. That is, the more Critical Questions are answered in the affirmative, the stronger the argument, whereas the more Critical Questions are answered in the negative, the weaker the argument (see Table 3.1 in Chapter 3).

First, let's consider CQ1. Could solar geoengineering get out of the control of (less than omnipotent) human beings? There is at least one sense in which deploying solar geoengineering techno-fixes, such as SAI and MCB, to counteract the negative impacts of climate change could get out of the control of less than all-powerful beings like us. Simply put, once we let the genies, be it aerosol particles in the case of SAI or sea salt particles in the case of MCB, out of the bottle, so to speak, there is no putting them back. As Alan Robock (2008, 17) expresses the concern:

> We don't know how quickly scientists and engineers could shut down a geoengineering system—or stem its effects—in the event of excessive climate cooling from large volcanic eruptions or other causes. *Once we put aerosols into the atmosphere, we cannot remove them* (emphasis added).

This sort of loss of control over solar geoengineering technologies could unfold in at least two different ways. In one scenario, we could get trapped in a cascade of negative impacts that we would be powerless to stop. That is, injecting aerosol particles into the stratosphere or seeding clouds with sea salt particles could trigger knock-on effects of unintended negative impacts that we would want to stop, but we would be powerless to do so. In another scenario, we could get trapped in a vicious loop from which we would be unable to escape. That is, injecting aerosol particles into the stratosphere or seeding clouds with sea salt particles could lead to (intended or unintended) positive impacts that we would want to maintain, but we would be powerless to maintain them without continuously injecting aerosol particles into the stratosphere or seeding clouds with sea salt particles. In either scenario, we could lose control over solar geoengineering technologies due to our limited power either to pull the plug on these technologies if things go wrong or to keep these technologies working continuously if things go well.

A related moral concern has to do with what has been dubbed "termination shock." According to Andy Parker and Peter Irvine (2018, 457), termination shock is "a rapid and substantial rise in global temperatures following a cessation of SRM deployment." In other words, given our limited power, we would have no way to sustain the deployment of solar geoengineering technologies continuously for long periods of time. But if the deployment of solar geoengineering technologies were to be successful, we would have to continue deploying these technologies in order to avoid the risk of termination shock. For if the deployment of these technologies were to be terminated abruptly, given that our ability to keep them operational for an extended period of time would be limited, it could trigger a fast and significant rewarming of the planet over which we would have no control at all. Alternatively, termination shock could be triggered by what Baum, Maher Jr., and Haqq-Misra (2013, 168) call "double catastrophe." In this scenario, we would be rendered powerless to maintain solar geoengineering interventions by "a catastrophic societal collapse." Such a catastrophic societal collapse "could be caused by a pandemic, nuclear war, or other global catastrophe" (Baum, Maher Jr., and Haqq-Misra 2013, 168).

From all of the above, it is reasonable to conclude that solar geoengineering could get out of the control of (less than omnipotent) human beings. Once we deploy solar geoengineering technologies, they could get out of our control because we would be unable to pull the plug on them immediately if we needed to or to sustain them continuously for long periods of time if we needed to (due to double catastrophe, for example), without risking termination shock. As Morrow (2020, 90) puts it, "the risk of this 'termination shock' remains a powerful reason not to use solar geoengineering as a stand-alone response to climate change." Accordingly, the answer to CQ1, as far as solar geoengineering is concerned, seems to be "yes" overall.

Next, let's consider CQ2. Could solar geoengineering be misunderstood or have unforeseen consequences for (less than omniscient) human beings? According to a Pew Research Center survey, 32 percent of US adults are *very concerned* that solar geoengineering technologies will be used before we fully understand how they affect the Earth's ecosystems, and 42 percent of US adults are *somewhat concerned* that solar geoengineering technologies will be used before we fully understand how they affect the Earth's ecosystems (Johnson and Kennedy 2021). There seems to be quite a lot that we, who are not omniscient as God is supposed to be, still do not understand about solar geoengineering. For instance, SAI was conceived by analogy with volcanic eruptions. More specifically, the 1991 volcanic eruption of Mount Pinatubo in the Philippines had a short-term cooling effect on the Earth's climate of approximately 0.6 degrees Celsius over a period of one year. However, the analogy between volcanic eruptions and SAI only goes so far. We do not know if injecting aerosol particles into the stratosphere would have exactly the same effects as a volcanic eruption. As Anni Määttänen et al. (2024, 2) put it, "volcanic eruptions and intentional injections would not have exactly the same impacts as *they are not perfect analogues*, differing for instance in terms of stratospheric aerosol size and composition or in the spatio-temporal characteristics of the injection" (emphasis added).

Moreover, we can only conduct SAI experiments on a rather small scale. For it to be a serious countermeasure against the disastrous impacts of climate change, however, solar geoengineering technologies would have to be deployed on a global scale. For this reason, the data we can gather from small-scale SAI experiments may or may not tell us much about what would happen when SAI is deployed on a global scale. As Määttänen et al. (2024, 2) put it, SAI field experiments "cannot inform us on the large-scale climate impacts of SAI nor do they allow for an experimental validation of climate responses to SAI." Likewise, Robock et al. (2010, 530) argue that "weather and climate variability preclude observation of the climate response without a large, decade-long forcing."

As for any unforeseen consequences of solar geoengineering, which could be harmful to us and the planet, there are many. For example, solar geoengineering could shift patterns of global atmospheric circulation in ways that we fail to foresee yet. This, in turn, could lead to more extreme weather events (United Nations Environment Programme 2023). In addition, solar geoengineering, specifically SAI, could alter global precipitation patterns in ways that we fail to foresee yet. This, in turn, could lead to reduced agricultural productivity (Lawrence et al. 2018), which would negatively affect the food supply, and thus the well-being, of many people around the world (Proctor et al. 2018). Zarnetske et al. (2021) have conducted a comprehensive survey of gaps in knowledge about the potential effects—both beneficial and harmful—of SAI. They conclude that "the effects of SAI on ecological systems are largely unknown" (Zarnetske et al. 2021, 2), including its effects on ecosystem services (e.g., fisheries), species interactions and food webs, droughts, migrations, hydrology, biogeochemistry, phenology, wildfires, species distributions, and more (Zarnetske et al. 2021). According to Robock et al. (2010, 530), we need "full-scale implementation" to understand the effects of SAI, but such "full-scale implementation could disrupt food production on a large scale." As for MCB, for all we know, it could actually increase the prevalence of high-altitude clouds, which would in turn warm the planet rather than cool it (Hoffmann and Feingold 2021).

From all of the above, it is reasonable to conclude that solar geoengineering could be misunderstood or have unforeseen consequences for (less than omniscient) human beings. Our current understanding of solar geoengineering technologies, such as SAI, is largely based on a likely inapt analogy with volcanic eruptions and is thus limited at best. There could also be numerous unforeseen, and potentially negative, consequences for us and the planet when solar geoengineering technologies are deployed on a global scale, given that what we do know about these technologies is based on small-scale, local field experiments. As Anna Lou Abatayo et al. (2020, 13393) put it, "the potential side effects [of solar geoengineering] include impacts on health and agriculture, as well as other adverse effects that are not yet fully understood." Accordingly, the answer to CQ2, as far as solar geoengineering is concerned, seems to be "yes" overall.

Let's move on to CQ3. Could solar geoengineering be misused or abused by (less than omnibenevolent) human beings? There are several ways in which solar geoengineering could be misused or abused by less than all-good human beings. That is, these technologies are dual use technologies insofar as they can be (mis)used (or abused) in ways other than their intended or designed use (see Chapter 3). Given the power that solar geoengineering technologies would grant to those who are able to develop and deploy them, and given the global impact that the deployment of such technologies would have over not only those who would have such technologies but also those who would not, it is easy to imagine how the former could misuse or abuse solar geoengineering technologies at the expense of the latter. In other words, the power imbalance that would be created by solar geoengineering technologies being developed and deployed by some but not others could be too tempting for less than all-good human beings not to exploit in the pursuit of power and profit. As Patrick Smith (2018, 357) puts it, "the ability to unilaterally deploy [solar geoengineering technologies] without input from relevant stakeholders characterizes the status quo. So, it seems like *abuse is a severe risk* at the moment without any power equalization" (emphasis added). The concern that any state actor that considers itself exceptionally endangered by climate change could unilaterally deploy solar geoengineering technologies, and thereby impose costs and risks on other state actors, is known as the "free-driver problem" (Weitzman 2015). Unilateral deployment decisions could lead to a world in which all nations, even those that were opposed to solar geoengineering, are locked into a geoengineered future.

A related moral concern is that of a technological arms race. The status quo Smith (2018, 357) alludes to in the quote above is that of environmental injustice. That is, affluent countries that have ample resources to develop and deploy solar geoengineering technologies are typically those that profited the most from industries and businesses largely responsible for greenhouse gas emissions, and thus largely responsible for anthropogenic climate change. On the other hand, less affluent countries that did not profit from environmental degradation and pollution typically have fewer resources to develop and deploy solar geoengineering technologies, even though they are the ones more likely to be harmed by solar geoengineering. According to Frank Biermann and Ina Möller (2019, 163):

> If deployed, climate engineering technologies are likely to have tremendous impacts on developing countries. And yet, *many climate engineering technologies could be deployed by a limited group of industrialised countries without global consent* (emphasis added).

So, a worry is that a technological arms race between affluent countries at the expense of less affluent countries would ensue. If such an arms race were to ensue, affluent countries could be tempted to research and develop their solar geoengineering technologies, including testing them on increasingly larger scales, even if it would cause harm to the environment and the people of less affluent countries.

The history of nuclear weapons testing can teach us a valuable lesson in this regard. For instance, between 1946 and 1958, the United States conducted sixty-seven nuclear tests in the Marshall Islands, according to the Atomic Heritage Foundation (2022). The Marshallese were forced to evacuate and relocate, which did not spare them from exposure to nuclear fallout. According to the Atomic Heritage Foundation (2022), "Many Marshallese have suffered from forced relocation, burns, birth defects, and cancers." Although nuclear testing in the Marshall Islands ended in 1958, "radiation related cancers and birth defects are a major problem faced by the Marshallese" (Atomic Heritage Foundation 2022). Furthermore, "Environmental issues such as rising sea levels, and degradation attributed to the nuclear testing, are also problems and often force residents to move between atolls" (Atomic Heritage Foundation

2022). It is not unreasonable to expect that affluent countries, like the United States, would be tempted to research and develop their solar geoengineering technologies even if it would cause harm to the environment and the people of less affluent countries. After all, as Clive Hamilton (2013, 107) points out, the US government has funded in the past studies aimed at "controlling the weather." In addition to some state actors researching and developing solar geoengineering technologies with no regard to the (potentially harmful) effects on other nations and peoples, there is also "a legitimate concern that research and testing of solar radiation management technologies would set the world on a slippery slope to ultimate deployment" (Hamilton 2013, 147), as state actors would ignore moral, social, legal, and other concerns about solar geoengineering technologies in the race to develop and deploy these technologies before other state actors do.

In that respect, another related moral concern is that of the weaponization of solar geoengineering technologies. According to Sovacool, Baum, and Low (2023), solar geoengineering technologies, such as SAI, may be too imprecise to directly weaponize, that is, to deploy in a hostile manner against a state actor or people specifically. However, the weaponization of solar geoengineering technologies could take other forms, such as "indirect weaponization" or "political weaponization" and "inadvertent weaponization" or "psychological weaponization" (Sovacool, Baum, and Low 2023, 12). As far as "indirect weaponization" or "political weaponization" is concerned, hostile deployment would not have to be real to provoke conflicts between state actors. For their study, Sovacool, Baum, and Low (2023, 12) interviewed one expert who put it this way:

> The political risks of deployment are very hard to predict, it depends on who is doing it, how it is being done, and who is deciding. It might not even need to be done to be a risk, to procreate a geopolitical conflict. A country could accuse their neighbour of doing strategic aerosol injection if their rainfall patterns have changed, or they saw some suspicious aircraft activity, or the signal-to-noise ratio for understanding atmospheric dynamics seemed different. It is hard to say with certainty and a short period of time whether one has done it, which means it could be used politically even without actually being used; a ploy, or an indirect or political weapon in that sense.

> *It doesn't have to be technically feasible to have a political effect as a signalling effect, to feed into a process of political weaponization.* The fact that many options are believed to be cheap, fast, and easy only makes this political threat more credible (emphasis added).

Like "indirect weaponization" or "political weaponization," "inadvertent weaponization" or "psychological weaponization" could also be perceived rather than real, but it could also be accidental. For their study, Sovacool, Baum, and Low (2023, 12) interviewed one expert who put it this way:

> It's possible that you could inadvertently weaponize these options. So you start a programme, and then there's a rapid warming that happens when you stop doing that stuff. In the termination-risk world, *you could accidentally weaponize something, because you stopped paying attention to it.* You could accidentally allow your virus to sneak out of your lab, because you weren't paying attention to that problem. So in that sense, it's not so much weaponization as just being an idiot, and then accidentally causing massive harm to the world. The perception of weaponization, though, is more interesting. *If we knew there was a fleet of Chinese or Russian aircraft flying around, modifying the climate, that's psychological weaponization.* It could make everyone feel vulnerable (emphasis added).

Recall, for instance, the panic that ensued in 2023 when a mysterious balloon was spotted over the United States. At the time, US officials claimed that it was a spy balloon or a surveillance device, whereas Chinese officials claimed that it was a weather balloon or a meteorological device. The balloon incident "strained the dynamic between two of the world's most powerful countries" and threatened "to derail any possibility of a diplomatic détente, with Beijing also accusing the United States of sending balloons to China" (Feng and Schapitl 2023).

In addition, there is also the risk that is referred to as "moral hazard." The worry is that the deployment of solar geoengineering technologies would give fossil fuel polluters continued license to pollute, thereby distracting from the difficult but required work of reducing greenhouse gas emissions. If it were to occur, moral hazard would be a misuse of solar geoengineering technologies because the deployment of such technologies is not meant to be a substitute for the reduction of greenhouse gas emissions. In other words, to use solar geoengineering technologies as a remedy for climate change, in lieu of reducing greenhouse gas emissions, is to misuse these technologies.

As Adam Corner and Nick Pidgeon (2014) explain, when it comes to solar geoengineering, the "moral hazard" problem poses risks at the individual, societal, and political levels. At the level of the individual, "the prospect of geoengineering could cause *individuals* to curb voluntary efforts to reduce carbon emissions" (Corner and Pidgeon 2014, 2). If that were the case, and solar "geoengineering were perceived by the public as offering a licence to continue carbon-intensive lifestyles, the 'rebound' [i.e., the idea that adopting an energy-saving technology can sometimes increase energy use, thereby undermining the overall impact of the technology itself] associated with geoengineering technologies could be even more significant" (Corner and Pidgeon 2014, 2–3). At the level of society, "Social norms specifying the urgency of preventative climate policies could shift" (Corner and Pidgeon 2014, 3). That is, "a particular individual may not believe that geoengineering policies represent a reason for acquiescing on low-carbon commitments, but may see a risk that others would modify their attitudes and behaviours" (Corner and Pidgeon 2014, 3). At the political level, state actors could divert resources from climate mitigation and adaptation efforts to the development and deployment of solar geoengineering technologies. All of these, namely, individual, social, and political moral hazards, are risky because solar geoengineering cannot be a substitute for climate mitigation and adaptation efforts. Even successful deployment of solar geoengineering technologies would not address the cause of the climate crisis, which is the high concentration of greenhouse gases in the Earth's atmosphere. As the authors of the Intergovernmental Panel on Climate Change (IPCC) Climate Change 2023 Synthesis Report (2023, 72) put it:

> Solar Radiation Modification (SRM) approaches, if they were to be implemented, introduce a widespread range of new risks to people and ecosystems, which are not well understood. SRM has the potential to offset warming within one or two decades and ameliorate some climate hazards but *would not restore climate to a previous state*, and substantial residual or overcompensating climate change would occur at regional and seasonal scales. Effects of SRM would depend on the specific approach used, and a sudden and sustained termination of SRM in a high CO_2 emissions scenario

> would cause rapid climate change. *SRM would not stop atmospheric CO2 concentrations from increasing nor reduce resulting ocean acidification under continued anthropogenic emissions.* Large uncertainties and knowledge gaps are associated with the potential of SRM approaches to reduce climate change risks. Lack of robust and formal SRM governance poses risks as deployment by a limited number of states could create international tensions (emphasis added).

In other words, as long as we continue to emit greenhouse gases, solar geoengineering technologies could amount to a temporary band-aid solution to the serious problem that is the climate crisis.

From all of the above, it is reasonable to conclude that solar geoengineering could be misused or abused by (less than omnibenevolent) human beings. Given that we are not morally perfect as God is supposed to be, the temptation to misuse or abuse solar geoengineering technologies could be too strong for

Table 5.1 Summary of the Answers (with Justifications) to the Critical Questions, CQ1–CQ3, of the "Playing God" Argument against Solar Geoengineering

Critical Question	Answer	Justification
CQ1: Could solar geoengineering get out of the control of (less than omnipotent) human beings?	yes	inability to pull the plug; inability for extended deployment (double collapse); risk of termination shock
CQ2: Could solar geoengineering be misunderstood or have unforeseen consequences for (less than omniscient) human beings?	yes	imperfect analogy with volcanic eruptions; unrepresentative local field experiments; global atmospheric circulation patterns could shift; global precipitation patterns could alter; more extreme weather events; reduced agricultural productivity
CQ3: Could solar geoengineering be misused or abused by (less than omnibenevolent) human beings?	yes	increased power imbalance between state actors; risk of unilateral deployment by some nations against others (free-driver problem); technological arms race (slippery slope); weaponization of solar geoengineering technologies; moral hazard

Source: Author

us to resist. In addition to simply misusing solar geoengineering technologies by pretending that they are a permanent "techno-fix" to the climate crisis, less than all-good peoples or nations could enter into technological arms races with other nations, unilaterally deploy solar geoengineering technologies to dominate other nations, and even use these technologies as weapons against other nations. Accordingly, the answer to CQ3, as far as solar geoengineering is concerned, seems to be "yes" overall.

As far as solar geoengineering is concerned, Critical Questions, CQ1–CQ3, seem to have affirmative answers (see Table 5.1 for a summary). This means that developing and deploying solar engineering technologies would indeed amount to "playing God" in terms of power (CQ1), knowledge (CQ2), and benevolence (CQ3). Therefore, a "playing God" argument against solar geoengineering is a strong argument (see Table 3.1 in Chapter 3). This argument can be summed up as follows.

Major Premise: We should not (morally speaking) "play God."

Conditional Premise: If solar geoengineering could get out of the control of humans, be misunderstood or have unforeseen consequences for humans, or be misused or abused by humans, then developing and deploying solar geoengineering techno-fixes to counteract climate change amount to "playing God."

Control Premise: Solar geoengineering could get out of the control of humans.

Understanding Premise: Solar geoengineering could be misunderstood or have unforeseen consequences for humans.

Use Premise: Solar geoengineering could be misused or abused by humans.

Conclusion: We should not (morally speaking) develop and deploy solar geoengineering techno-fixes to counteract climate change.

This strong "playing God" argument against solar geoengineering points to several moral problems with solar geoengineering technologies as techno-fixes for the climate crisis, namely, that they could get out of control post-deployment, that the ways they would work could be misunderstood, that they could have unintended long-term consequences for our environment and the planet, and that they could be misused or abused by nations and state actors.

Like all "playing God" arguments, this one against solar geoengineering is a defeasible argument. As we have seen in Chapter 3, conclusions of defeasible arguments can always be defeated by additional evidence that is not cited in their premises. In other words, this argument is not the final word about whether or how we should develop and deploy solar geoengineering technologies to address the problem of climate change. One obvious way to defeat this "playing God" argument against solar geoengineering is to argue that the potential benefits of solar geoengineering would outweigh the perils of loss of control, misunderstanding, unforeseen consequences, misuse, and abuse. Still, this strong "playing God" argument against solar geoengineering urges us to think carefully about the ways in which the development and deployment of solar engineering technologies could go seriously morally wrong. So, until we can make sure that we are able to control solar engineering technologies, understand them and their effects, and use them benevolently, we have a strong case against developing and deploying solar engineering technologies.

As we have seen in Chapter 3, our answers to the Critical Questions, CQ1–CQ3, will tell us what issues need to be addressed before the new or emerging technology in question can be developed, deployed, or used responsibly, that is, in accordance with our human limitations and without pretending that we are all-powerful, all-knowing, and all-good. In general, an affirmative answer to CQ1 suggests that the new or emerging technology in question could get out of the control of (less than omnipotent) human beings. Of course, we could never have perfect control over a new or emerging technology because we are not all-powerful as God is supposed to be. However, we can try to enhance our control over a new or emerging technology by analyzing risks and designing for safety. Accordingly, a "yes" to CQ1 suggests that we should direct our efforts toward risk analysis and safety. That is, we should conduct extensive risk analyses and safety testing, as well as design safety mechanisms informed by the results of these risk assessments and safety tests, to make sure that the new or emerging technology in question would be controllable by us before it can be deployed or used responsibly, that is, in accordance with our human limitations and without pretending that we are all-powerful.

An affirmative answer to CQ2 suggests that the new or emerging technology in question could be misunderstood or have unforeseen consequences for (less than omniscient) human beings. Of course, we could never have perfect understanding of a new or emerging technology and know all of its potential impacts because we are not all-knowing as God is supposed to be. However, we can try to improve our knowledge and understanding of a new or emerging technology through further research. Accordingly, a "yes" to CQ2 suggests that we should direct our efforts toward research. That is, we should conduct extensive experiments and simulations to learn as much as we can about the new or emerging technology in question before it can be deployed or used responsibly, that is, in accordance with our human limitations and without pretending that we are all-knowing.

An affirmative answer to CQ3 suggests that the new or emerging technology in question could be misused or abused by (less than omnibenevolent) human beings. Of course, we could never deploy or use a new or emerging technology in perfectly good ways because we are not all-good as God is supposed to be. However, we can try to prevent misuse and abuse by enacting regulations, policies, and laws with robust enforcement mechanisms. Accordingly, a "yes" to CQ3 suggests that we should direct our efforts toward policy. That is, we should conduct extensive deliberations, which would include all stakeholders, aimed at drafting regulations or legislating laws pertaining to the new or emerging technology in question before it can be deployed or used responsibly, that is, in accordance with our human limitations and without pretending that we are all-good.

Now, as far as the "playing God" argument against solar geoengineering is concerned, we have answered Critical Questions, CQ1–CQ3, in the affirmative. This gives us good reasons to be concerned about solar geoengineering getting out of control, being misunderstood or having unforeseen consequences, and being misused or abused. Having good reasons to believe this would, in turn, help us to direct our governance efforts as far as solar geoengineering is concerned. First, our affirmative answer to CQ1 suggests that we should direct our governance efforts toward conducting risk assessments and safety tests, as well as designing safety mechanisms informed by the results of these risk assessments and safety tests, for solar geoengineering technologies. Second, our affirmative answer to CQ2 suggests that we should direct our governance efforts toward further research on solar geoengineering interventions, such as SAI, and their potentially harmful impacts on the environment and the climate. Third, our affirmative answer to CQ3 suggests that we should direct our governance efforts toward crafting policies for the development and deployment of solar geoengineering technologies, so as to prevent misuse and abuse by state actors. In that respect, the "playing God" argument against solar geoengineering, along with our answers to the Critical Questions, CQ1–CQ3, may provide some general guidance to solar geoengineering governance.[3]

To sum up, here is how our answers to Critical Questions, CQ1–CQ3, can provide some general guidance for approaching the governance of solar geoengineering.

CQ1: Could solar geoengineering get out of the control of (less than omnipotent) human beings?

- Yes: attempt to enhance control by focusing on safety through risk assessment, safety testing, etc.

CQ2: Could solar geoengineering be misunderstood or have unforeseen consequences for (less than omniscient) human beings?

- Yes: attempt to increase knowledge and understanding of consequences by focusing on research through experimentation, simulation, etc.

CQ3: Could solar geoengineering be misused or abused by (less than omnibenevolent) human beings?

- Yes: attempt to prevent misuse or abuse by focusing on policy through regulation, legislation, enforcement, etc.

Notes

1 There are other technologies that broadly fall under the umbrella of geoengineering or solar geoengineering. In this chapter, I focus on SRM in particular.

2 Cf. Laura Hartman (2017) who argues that the "playing God" critique is not "a moral argument" (317) at all but rather "a rhetorical move that privileges evocative analogy over logical argument" (315). For Massimiliano Simons (2022), too, "playing God" is a symbolic critique rather than a logical argument.

3 Recall that the answers to the Critical Questions, CQ1–CQ3, of a "playing God" argument with respect to a new or emerging technology are meant to provide general guidance for approaching the governance of that technology. They are not supposed to constitute concrete governance recommendations or serve as a comprehensive governance proposal. On concrete governance proposals for the governance of solar geoengineering, such as the Oxford Principles (Rayner et al. 2013) and Tollgate Principles, see Stephen Gardiner and Augustine Fragnière (2018).

6

Are We "Playing God" with Artificial Intelligence?

In March 2023, the Future of Life Institute published an open letter, which was signed by several scientists, engineers, and leaders in the digital tech industry, entitled "Pause Giant AI Experiments." The open letter states that "AI [i.e., Artificial Intelligence] systems with human-competitive intelligence can pose profound risks to society and humanity" (Krakovna et al. 2023). Despite these risks, the letter continues, digital tech companies, such as OpenAI and Google's DeepMind, are "locked in an out-of-control race to develop and deploy ever more powerful digital minds that no one—not even their creators—can understand, predict, or reliably control" (Krakovna et al. 2023).

Recall the fundamental questions about the limits of our power to control our environment, our knowledge of our world and our place in it, and our corruptible nature or moral character from Chapter 1.

- Can we control it?
- Do we understand it?
- Will we be corrupted by it?

The open letter seems to express concerns about our ability to control, understand, and reign in AI technologies. The authors of the open letter seem to think that we cannot "reliably control" these "powerful digital minds." So, this is their negative answer to the question, "Can we control it?" The authors of the open letter also seem to think that we cannot "understand" or "predict" these "powerful digital minds." So, this is their negative answer to the question, "Do we understand it?" Finally, the authors of the open letter express the moral concerns that animate the question "Will we be corrupted by it?" by raising the following questions:

> Should we let machines flood our *information channels* with propaganda and untruth? Should we automate away all the *jobs*, including the *fulfilling ones*? Should we develop nonhuman minds that might eventually outnumber, *outsmart*, obsolete and replace us? Should we risk *loss of control of our civilization*? (emphasis added)

In other words, the authors of the open letter are concerned about the potentially perversive effects that AI technologies would have on us and how we relate to other people, including our work lives ("jobs"), our social lives ("information channels"), and our political lives ("our civilization") as we continue to design, develop, deploy, and use AI technologies. One year after the open letter was signed, the AI arms race is speeding up, not slowing down (Aguirre 2024). This is a testament to how the promise of new technology can corrupt us. That is, the alluring promise of machines doing our work for us and making money for us is simply too irresistible for less than all-good beings like us. In that sense, the authors of the open letter that calls for a pause on AI experiments, such as GPT-4, are effectively making a "playing God" argument, even though they do not use the phrase "playing God" explicitly.

Indeed, one of the AI developers to whom the open letter was addressed, namely, Ilya Sutskever, who worked at OpenAI when the tech company was developing its flagship AI chatbot, ChatGPT, acknowledges the "playing God" argument against AI. In an interview, Sutskever put it this way (Varga 2023):

> *Playing God*, scientists have been accused of *playing God* for a while, but there is a real sense in which we are creating something very different from anything we've created so far. Yeah, I mean, *we definitely will be able to create completely autonomous beings with their own goals* (emphasis added).

In other words, for Sutskever, creating autonomous beings that have their own goals amounts to "playing God."

So, let's consider carefully the "playing God" argument against AI. Using my argumentation scheme for "playing God" arguments (see Chapter 3), such a "playing God" argument against AI can be stated as follows.

> *Major Premise*: We should not (morally speaking) "play God."
> *Minor Premise*: Developing and deploying AI technologies amount to "playing God."

Conclusion: We should not (morally speaking) develop and deploy AI technologies.

Before the conclusion of this argument can be accepted or rejected, we should evaluate this argument by asking the Critical Questions that accompany the argumentation scheme for "playing God" arguments. These questions are as follows.

CQ1: Could AI get out of the control of (less than omnipotent) human beings?
CQ2: Could AI be misunderstood or have unforeseen consequences for (less than omniscient) human beings?
CQ3: Could AI be misused or abused by (less than omnibenevolent) human beings?

If the answer to all of these Critical Questions is "yes," then it means that developing and deploying AI technologies would indeed amount to "playing God" in terms of power, knowledge, and benevolence. Given that we should not pretend to be all-powerful, all-knowing, and all-good as God is supposed to be, we could then conclude that we should not develop and deploy AI technologies. In that case, the "playing God" argument against AI would be a strong argument. On the other hand, if the answer to all of these Critical Questions is "no," then it means that developing and deploying AI technologies would not amount to "playing God." In that case, the conclusion that we should not develop and deploy AI technologies would not be warranted and the "playing God" argument against AI would be a weak argument. Any other combination of "yes" and "no" answers to the Critical Questions, CQ1–CQ3, would make the "playing God" argument against AI stronger or weaker. That is, the more Critical Questions are answered in the affirmative, the stronger the argument, whereas the more Critical Questions are answered in the negative, the weaker the argument (see Table 3.1 in Chapter 3).

First, let's consider CQ1. Could AI get out of the control of (less than omnipotent) human beings? As we have seen, the authors of the open letter think that we would not be able to "reliably control" AI (Krakovna et al. 2023). The control problem is a well-known problem in the field of Artificial Intelligence. According to Paula Boddington (2017, 29), the more powerful AI is, as a general rule, the less we are going to be able to control it. As Stuart Russell (2022, 20) explains, "Once AI systems move out of the laboratory (or artificially defined environments such as the simulated chessboard) and into the real world, there is very little chance that we can specify our objectives completely and correctly in such a way that the pursuit of those objectives by more capable machines is guaranteed to result in beneficial outcomes for humans." As far as AI technologies are concerned, then, the control problem is an instance of the general concern about beings with limited power, such as human beings, losing control over technology, which is precisely the concern addressed by CQ1.

In that respect, it may be useful to distinguish between different kinds of AI in terms of the capabilities of these technologies. First, so-called "weak AI" or "narrow AI" refers to computing machines or programs that simulate or mimic intelligence capabilities without necessarily being intelligent themselves. For the most part, this is the kind of AI we currently have. The so-called Generative AI (GenAI) technologies that are all the rage nowadays, such as Large Language Models (LLMs) and diffusion models, are trained on large amounts of data, which are typically scraped from websites and apps, and then processed by means of machine learning algorithms, such that they are able to produce outputs in various formats, including text, audio, and video. Second, so-called "strong AI" or "broad AI" refers to computing machines or programs that have intelligence capabilities in the form of understanding or consciousness. No such AI systems exist yet. Third, so-called "Artificial General Intelligence" or AGI refers to computing machines or programs that exhibit the full range of intelligence capabilities that human beings possess. Again, no such AI systems exist yet. Finally, so-called "superintelligence" refers to computing machines or programs that surpass human intelligence. This is the kind of AI usually depicted in science fiction books, such as HAL 9000 in Arthur C. Clarke's *Space Odyssey* series, and movies, such as Skynet in the *Terminator* series. Needless to say, no such AI systems exist yet.

Now, it seems clear that we could lose control over AGI and superintelligence. After all, such AI systems would be as smart or even smarter than humans. Moreover, having the full range of intelligence capabilities that we have, including understanding and consciousness, they would have their own thoughts, ideas, wants, desires, goals, and aspirations. In that respect, we could have very little to no control over what such AI systems would think or want. As Alan Turing once put it during a BBC broadcast, "It seems probable that once the machine thinking method had started, it would not take long to outstrip our feeble powers. [. . .] At some stage therefore we should have to expect the machines to take control" (quoted in Russell 2022, 20). Nevertheless, there is a sense in which we could lose control over even weak or narrow AI, including GenAI, which is the only kind of AI we have so far.

Our loss of control over weak or narrow AI could be gradual. Contrary to a drastic usurpation of power by AGI or superintelligence, this loss of control over AI could be barely noticeable to us. In such a scenario, AI could steadily rob us of our power and control "by stealthily taking away decision-making power app by app, or by diminishing our capacities as we rely more and more on machines to do our thinking for us" (Boddington 2023, 26–7). For instance, as AI systems play an increasingly larger role in decision-making, such as approving or denying loans, medical claims, insurance coverage, college applications, and so on, humans would gradually lose control over making such decisions. As Luciano Floridi (2023) warns, "automation bias," or the "perceived mechanistic objectivity associated with computer-generated analytics" (97) already leads us to uncritically accept machine-generated decisions with an uncritical shrug of the shoulders and "the computer said so" (110). If we continue to outsource decision-making to AI systems, and accept "the computer said so" as a justification without further reflection, we will continue to lose more of our decision-making power. John Danaher (2016, 246) refers to this sort of loss of human control to AI as "*the threat of algocracy*—a situation in which algorithm-based systems structure and constrain the opportunities for human participation in, and comprehension of, public decision-making" (emphasis added).

Another sense in which current AI technologies, like GenAI, are getting out of our control has to do with digital pollution. While AI technologies do have a massive carbon footprint (more on that below), they are also responsible for digital or internet pollution, which is getting out of control. The problem of digital pollution created by GenAI, which is often referred to as "AI slop," is exacerbated by the fact that the digital pollution, that is, poor quality synthetic data created by GenAI that outnumbers good quality data created by humans, is then scraped off the internet and used by tech companies to train the very GenAI models that create more synthetic data of poor quality. This vicious cycle of synthetic data generation pollutes the internet with AI-generated text, images, and videos of poor quality. As Erik Hoel (2024) puts it, "Increasingly, mounds of synthetic A.I.-generated outputs drift across our feeds and our searches. The stakes go far beyond what's on our screens. The entire culture is becoming affected by A.I.'s runoff, an insidious creep into our most important institutions [including academic journals of science and medicine]."

Even tech companies are concerned about the quality of the data that are being scraped off the internet. For it turns out that training GenAI models on synthetic data created by other GenAI models could lead to so-called "model collapse." According to Ilia Shumailov et al. (2024, 2), model collapse is "a degenerative process whereby, over time, models forget the true underlying data distribution, even in the absence of a shift in the distribution over time." This degradation in synthetic data and generative output quality is "inevitable" (Shumailov et al. 2024, 2) if GenAI models are trained on synthetic data that are generated by other GenAI models. As Shumailov et al. (2024, 2) put it, "the use of LLMs at scale to publish content on the Internet will *pollute* the collection of data to train them" (emphasis added).[1]

From all of the above, it is reasonable to conclude that AI could get out of the control of (less than omnipotent) beings like us. As AI systems become more capable, they could rob us of our power and control over our decision-making and thinking gradually and with little to no resistance from us. AI systems with intelligence capabilities like ours or superior to ours, that is, AGI and superintelligence, could overpower us even if we were to resist them. What's more, the digital pollution that extant AI systems create, namely, AI slop, is already getting out of our control. Accordingly, the answer to CQ1, as far as AI is concerned, seems to be "yes" overall.

Next, let's consider CQ2. Could AI be misunderstood or have unforeseen consequences for (less than omniscient) human beings? There are good reasons to believe that the answer to CQ2 is "yes." For AI systems present a fundamental problem that significantly limits our understanding of these technologies. It is the so-called "black box problem." Take, for example, the system known as AlphaFold. AlphaFold was developed by DeepMind, a subsidiary of Google's parent company, Alphabet, to predict a protein's three-dimensional structure from its amino acid sequence. However, scientists cannot explain the system's predictions. In other words, they know what goes into the system (i.e., the input) and they know what comes out of the system (i.e., the output), but they do not know what happens in between. That is because AlphaFold is a deep learning artificial neural network (DNN). DNNs are black boxes. They are comprised of multiple layers, including nodes and edges, of mathematical relations, which makes it practically impossible for us to explain how they arrive at their predictions and decisions. As far as black box AI systems are concerned, "it [is] difficult to guess how they make decisions or why they create a certain output" (Garrido-Rodriguez et al. 2024, 16). This is the black box problem (Hassija et al. 2024).[2] Because of the black box problem with AI systems like Deep Neural Networks (DNNs), some researchers call for rethinking the ways in which AI systems are designed and developed. They call for making AI systems explainable and their outputs interpretable by humans. Explainable AI or "XAI methods attempt to explain black box systems (e.g., DNN) by building a second 'explanation' model" (Fleischer 2022, 535). Nevertheless, as long as we do not know what goes on under the hood, so to speak, there could be a lot that we do not understand or misunderstand about how AI systems generate the outputs that they do.

As for any consequences that would be unforeseen for less than omniscient beings like us, AI technologies could have quite a few, including uncertain environmental impacts. For instance, we already do know that AI has a massive carbon footprint (Floridi 2023, 183). That is because the vast amounts of data consumed by AI systems need to be stored in large data centers where servers that require lots of energy to operate are housed. These data centers, and the AI systems that consume the data stored there, generate greenhouse gas emissions. But we do not know exactly how massive the carbon footprint of AI is or will be as "the out-of-control and unregulated race to develop and deploy increasingly powerful [AI] systems" (Aguirre 2024) is well underway. So, the "challenge [. . .] is assessing and accounting for the carbon footprint of the technology itself" (Floridi 2023, 186). According to a recent estimate by the International Energy Agency (IEA), the energy consumption of data centers could double by 2026 (IEA 2024).

In addition to energy, data centers also consume vast amounts of potable water. As AI applications run on them, servers consume a lot of energy and generate heat as a result, which is why water is needed to keep the servers cool. This is another uncertain consequence of the development and deployment of AI systems. That is, we do not know what effects these technologies will have on the water supply that we all depend on. But we are already beginning to see some alarming signs. For example, one study found that OpenAI's ChatGPT consumes 500 milliliters of water, which is roughly equivalent to one sixteen-ounce water bottle, for every five to fifty prompts (Li et al. 2023). Some American towns are already pushing back against tech companies that seek to build data centers in their area. For example, "In July 2022, the month before OpenAI says it completed its training of GPT-4, Microsoft pumped in about 11.5 million gallons of water to its cluster of Iowa data centers, according to the West Des Moines Water Works. That amounted to about 6% of all the water used in the district, which also supplies drinking water to the city's residents" (O'Brien and Fingerhut 2023). Subsequently, the city of West Des Moines issued a statement that it "will only consider future data center projects" that can "demonstrate and implement technology to significantly reduce peak water usage from the current levels" (O'Brien and Fingerhut 2023).

AI not only consumes energy and water in vast quantities but also contributes to environmental pollution through the production of hardware for AI and data centers. Producing the hardware on which AI systems run requires mining minerals, such as silicon, germanium, gallium, boron, and phosphorus. The mining of these minerals contributes to environmental degradation and water pollution (Mishra 2015). The extraction of precious minerals that are required for producing the hardware on which AI systems run could also lead to geopolitical conflicts that we fail to foresee yet as the demand for such minerals is only expected to rise. Indeed, the United States, the European Union, and China are already locked in a (thus far unarmed) "race for Africa's critical mineral resources" (Boafo et al. 2024). As critical minerals become more difficult to find and harder to extract, while the demand for them continues to increase substantially, the economic conflict between those vying for control over Africa's critical mineral resources could easily turn into an armed conflict.

From all of the above, it is reasonable to conclude that AI could be misunderstood or have unforeseen consequences for (less than omniscient) human beings. AI black boxes are opaque to us insofar as we do not understand how they generate the outputs that they do. The carbon and water footprints of AI systems are uncertain, although they are increasing as the training of AI systems on massive amounts of data stored in large data centers demands more computing power and energy, which could have unforeseen consequences for us and our environment. The demand for and extraction of precious minerals to build the hardware for AI, as well as data centers to train more AI models, could lead to heretofore unforeseen geopolitical (possibly armed) conflicts. Accordingly, the answer to CQ2, as far as AI is concerned, seems to be "yes" overall.

Let's move on to CQ3. Could AI be misused or abused by (less than omnibenevolent) human beings? There are several ways in which AI technologies could be misused or abused by less than all-good beings like us. In fact, bad actors already use AI to spread misinformation, commit fraud, perpetrate scams, manipulate consumers and voters, influence stock markets, and more. As Kate Crawford once said, AI's potential for abuse is "a fascist's dream" (Solon 2017). This is because AI technologies are dual use technologies insofar as they can be (mis)used (or abused) in ways other than their intended or designed use (see Chapter 3). For instance, since ChatGPT was released in 2022, students have been (mis)using it to cheat on their school assignments, thereby creating a "massive plagiarism problem" (Gecker 2023). Even lawyers have been caught cheating with ChatGPT after they submitted a legal brief that contained fake cases generated by the chatbot (Acres 2023).

As we have seen in Chapter 3, the spread of misinformation, disinformation, and fake news online is already getting out of control. AI technologies can be used to scale up the spread of misinformation, disinformation, and fake news online with an ease and speed that bad actors never had before. According to Ilya Sutskever, formerly of OpenAI, because of AI, the "problem of fake news is going to be a million times worse, cyberattacks will become much more extreme. We will have totally automated weapons. I think AI has the potential to create infinitely stable dictatorships" (Varga 2023).[3]

As far as AI-enabled misinformation is concerned, AI technologies can be used to produce what are known as "deepfakes," that is, AI-generated, synthetic digital content or media, including audio, images, and videos, which depict events that never happened or feature people (often famous people) saying things they never said and doing things they never did in real life. As Daniele Battista (2024, 14) observes, deepfakes "threaten national security by disseminating propaganda and potential interference in elections." Recent examples include deepfake phone calls of US President Joe Biden urging registered Democrats in New Hampshire to not vote in the 2024 US presidential election (Bond 2024), deepfake images of Taylor Swift and some so-called "Swifties" (i.e., Taylor Swift fans) endorsing US presidential candidate Donald Trump (Wong and Lee 2024), and deepfake videos of Indian Prime Minister Narendra Modi giving a personalized greeting for the Hindu holiday of Diwali (Verma and Zakrzewski 2024). According to Battista (2024, 13–14), "the potential ability of deepfakes [. . .] to influence democratic elections [goes] far beyond the data analysis techniques previously associated with cases like Cambridge Analytica" (see Chapter 3).[4]

A particularly morally reprehensible use of AI to create deepfakes of unsuspecting people is so-called "synthetic porn" or "deepfake pornography." In a *Wired* article, titled "Deepfake Porn Is Out of Control," Matt Burgess (2023) writes:

> Since deepfakes emerged half a decade ago, the technology has consistently been used *to abuse and harass women*—using machine learning to morph someone's head into pornography without their permission. Now *the number of nonconsensual deepfake porn videos is growing at an exponential rate*, fueled by the advancement of AI technologies and an expanding deepfake ecosystem (emphasis added).

Even teenagers can now become victims of deepfake porn. For example, a high school student from Texas became a victim of deepfake porn when she was fourteen years old. "In October 2023, a student took her Instagram photos and plugged them into artificial intelligence, creating fake nude images of the 14-year-old girl" (Liou 2024). Given the absence of laws regulating AI-generated synthetic media at this time, some victims have become advocates for federal regulations, such as the so-called "Disrupt Explicit Forged Images and Non-Consensual Edits" (Defiance) Act, which would allow victims to sue their abusers (Tenbarge 2024). In that respect, it seems clear that AI technologies already have a corrupting influence on us, insofar as we (mis)use them to create synthetic media that could be used to harass or hurt other people.

AI technologies can be used to create synthetic media to not only harass or abuse people but also commit fraud. GenAI, in particular, has given rise to a new form of identity theft, namely, Synthetic Identity Fraud (SIF). As Muhammad Hanif (2024) explains, "Traditionally, ID theft relies on people's stolen information. In contrast, synthetic identity fraud involves creating an entirely new person who doesn't exist legally by combining stolen, manipulated and fictitious personally identifiable information along with bogus facial imprints" (emphasis added). Cybercriminals then create synthetic imagery and fraudulent documents to match the synthetic identity, thereby fooling extant ID verification and biometric authentication systems. According to the Federal Reserve (Federal Reserve Banks 2019), SIF is the fastest growing type of financial crime in the United States, and it accounts for billions of dollars in losses annually.

As far as AI-enabled crime goes, Floridi (2023, 116) identifies "five crime areas potentially affected by AIC [i.e., AI Crime]":

1) commerce, financial markets, and insolvency (including trading, bankruptcy);
2) harmful or dangerous drugs (including illicit goods);
3) offences against the person (including homicide, murder, manslaughter, harassment, stalking, torture);
4) sexual offences (including rape, sexual assault); and
5) theft and fraud, and forgery and personation.

An example of (1) is the use of AI systems that can learn techniques of financial market manipulation, such as order-book spoofing (Floridi 2023, 121). An example of (2) is the use of Unmanned Underwater Vehicles (UUVs) for drug trafficking (Floridi 2023, 124).[5] An example of (3) is the use of GenAI, particularly chatbots like ChatGPT and the like, to harass people (Floridi 2023, 124–5). An example of (4) is the use of physical sex robots (or sexbots) to simulate sexual offenses, such as rape, which could promote "sexual objectification or sexualized abuse and violence" as well as heighten "sexual desire for sexual offences" (Floridi 2023, 127). An example of (5) is the use of AI to automate and scale up cyberattacks, such as denial-of-service attacks (i.e., flooding a server with traffic to prevent users from accessing connected services and sites), social engineering, such as phishing (i.e., attempting to trick victims into giving away personal information by means of deceitful emails, voice, or text messages), and other forms of fraud and cybercrime (Floridi 2023, 129–30).

There are many more ways in which AI can be misused or abused. Slattery et al. (2024) have created the MIT AI Risk Repository, which identifies over sixteen hundred risks from AI. The AI risks are classified by domains as follows:

1. discrimination and toxicity
    1.1. unfair discrimination and misrepresentation
        1.1.1. biased output
        1.1.2. harmful decisions
    1.2. exposure to toxic content
        1.2.1. content that promotes violence, hate, and unlawful activities
        1.2.2. hazardous or misleading high-risk advice
        1.2.3. offensive explicit material
    1.3. unequal performance across groups
        1.3.1. alienation, frustration, and exclusion
        1.3.2. lower performance for some languages
        1.3.3. reduced ability to benefit from the system
2. privacy and security
    2.1. compromise of privacy by obtaining, leaking, or correctly inferring sensitive information
        2.1.1. adversaries gain knowledge of specific private records used to train an AI
        2.1.2. maliciously designed and used AI exploits users trust
        2.1.3. models inadvertently reproduce sensitive personal information

- 2.1.4. models reproduce classified intellectual property
- 2.1.5. prompt-based privacy attack
- 2.2. AI system security vulnerabilities and attacks
  - 2.2.1. direct manipulation
  - 2.2.2. external tool and API integration
  - 2.2.3. physical and network
  - 2.2.4. toolchain and dependency vulnerabilities
- 3. misinformation
  - 3.1. false or misleading information
    - 3.1.1. common sources of misinformation
    - 3.1.2. physical, emotional, and material harm
    - 3.1.3. undermining of autonomy
  - 3.2. pollution of information ecosystem and loss of consensus reality
    - 3.2.1. hinder constructive dialogue
    - 3.2.2. splintering of shared reality
    - 3.2.3. undermine health of democracy
- 4. malicious actors and misuse
  - 4.1. disinformation, surveillance, and influence at scale
    - 4.1.1. advanced AI in phishing schemes
    - 4.1.2. illegitimate domestic surveillance
    - 4.1.3. large-scale manipulation and propaganda
    - 4.1.4. mass gathering of personal data
    - 4.1.5. microtargeting
  - 4.2. cyberattacks, weapon development or use, and mass harm
    - 4.2.1. development and application of weapons
    - 4.2.2. large-scale harm
    - 4.2.3. Lethal Autonomous Weapons
    - 4.2.4. malicious malware
  - 4.3. fraud, scams, and targeted manipulation
    - 4.3.1. cheating
    - 4.3.2. deepfakes
    - 4.3.3. enhance dishonest schemes
- 5. human-computer interaction
  - 5.1. overreliance and unsafe use

        - 5.1.1. exploitation of trust
        - 5.1.2. miscalibrated trust
        - 5.1.3. weakened social ties
    - 5.2. loss of human agency and autonomy
        - 5.2.1. cognitive enfeeblement
        - 5.2.2. helplessness and job displacement
        - 5.2.3. limit personal growth
- 6. socioeconomic and environmental harms
    - 6.1. power centralization and unfair distribution of benefits
        - 6.1.1. censorship, oppression, and surveillance
        - 6.1.2. denial of access to critical services
        - 6.1.3. entrenchment of existing inequalities
    - 6.2. increased inequality and decline in employment quality
        - 6.2.1. decreased job quality and security
        - 6.2.2. exploitative dependencies
        - 6.2.3. mass unemployment
        - 6.2.4. worsen existing inequalities
    - 6.3. economic and cultural devaluation of human effort
        - 6.3.1. expropriation of cultural value
        - 6.3.2. extraction of copyright-protected works
        - 6.3.3. homogenization of cultural experiences
        - 6.3.4. stymie creativity
    - 6.4. competitive dynamics
        - 6.4.1. arms race
        - 6.4.2. cutting of safety corners
        - 6.4.3. deprioritization of long-term societal wellbeing
        - 6.4.4. widespread externalities
    - 6.5. governance failure
        - 6.5.1. difficulty of influencing developers and deployers
        - 6.5.2. rapid pace of AI development
        - 6.5.3. unclear liability
    - 6.6. environmental harm
        - 6.6.1. emissions from AI-enabled applications
        - 6.6.2. model hardware and data centers
        - 6.6.3. model training

7. AI system safety, failures and limitations
    7.1. AI pursuing its own goals in conflict with human goals or values
        7.1.1. acquisition and use of dangerous capabilities
        7.1.2. resist control
        7.1.3. technical challenges
    7.2. AI possessing dangerous capabilities that could cause mass harm
        7.2.1. cyber-offense skills
        7.2.2. evasive skills
        7.2.3. manipulation and persuasion
        7.2.4. political strategy and knowledge of social dynamics
        7.2.5. self-proliferation
        7.2.6. strategic planning
    7.3. lack of capability or robustness
        7.3.1. lack of inherent capability or skill
        7.3.2. out of distribution situations
        7.3.3. oversights, undetected bugs, or errors in the design process
        7.3.4. unusual changes or perturbations in input data
    7.4. lack of transparency or interpretability
        7.4.1. frustrate achievement of auditing standards
        7.4.2. lack of trust and confidence in systems results
    7.5. AI welfare and rights
        7.5.1. mistreatment or harm of AI systems[6]

On 2.2, that is, security vulnerabilities of AI systems, in particular, see also the National Institute of Standards and Technology's taxonomy of attacks on AI systems (Vassilev et al. 2024).

From all of the above, it is reasonable to conclude that AI could be misused or abused by (less than omnibenevolent) human beings. Given that we are not morally perfect as God is supposed to be, the temptation to misuse or abuse AI could be too hard for us to resist. In fact, GenAI technologies are already being used to spread misinformation, harass and abuse people, especially women, and commit fraud and other types of crime. AI could also be weaponized in the form of Lethal Autonomous Weapon Systems (LAWS).[7] Accordingly, the answer to CQ3, as far as AI is concerned, seems to be "yes" overall.

As far as AI is concerned, Critical Questions, CQ1–CQ3, seem to have affirmative answers (see Table 6.1 for a summary). This means that developing and deploying AI technologies would indeed amount to "playing God" in terms of power (CQ1), knowledge (CQ2), and benevolence (CQ3). Therefore, a "playing God" argument against AI is a strong argument (see Table 3.1 in Chapter 3). This argument can be summed up as follows.

Major Premise: We should not (morally speaking) "play God."
Conditional Premise: If AI could get out of the control of humans, be misunderstood or have unforeseen consequences for humans, or be misused or abused by humans, then developing and deploying AI amount to "playing God."
Control Premise: AI could get out of the control of humans.
Understanding Premise: AI could be misunderstood or have unforeseen consequences for humans.
Use Premise: AI could be misused or abused by humans.
Conclusion: We should not (morally speaking) develop and deploy AI.

Table 6.1 Summary of the Answers (with Justifications) to the Critical Questions, CQ1–CQ3, of the "Playing God" Argument against AI

Critical Question	Answer	Justification
CQ1: Could AI get out of the control of (less than omnipotent) human beings?	yes	the control problem; usurpation by AGI or superintelligence; loss of power and control through outsourcing decision-making and thinking to AI; out-of-control digital pollution; model collapse
CQ2: Could AI be misunderstood or have unforeseen consequences for (less than omniscient) human beings?	yes	the black box problem; uncertain carbon footprint; unknown water footprint; uncertain environmental impacts of energy and water consumption of AI and data centers; unforeseen geopolitical (armed) conflicts
CQ3: Could AI be misused or abused by (less than omnibenevolent) human beings?	yes	cheating and plagiarism; manipulation of voters or consumers with AI-generated misinformation; fraud by means of AI-generated deepfakes, voice clones, and SIF; abuse and harassment by means of AI-generated, synthetic media (e.g., deepfake porn); AI-enabled crime; weaponization of AI systems

Source: Author

This strong "playing God" argument against AI points to several moral problems with AI technologies, namely, that they could get out of control before and after deployment, that the way they work could be misunderstood, that they could have unintended long-term consequences for us and the planet, and that they could be misused or abused by people.

Like all "playing God" arguments, this one against AI is a defeasible argument. As we have seen in Chapter 3, conclusions of defeasible arguments can always be defeated by additional evidence that is not cited in their premises. In other words, this argument is not the final word about whether or how we should develop and deploy AI technologies. One obvious way to defeat this "playing God" argument against AI is to argue that the potential benefits of AI would outweigh the perils of loss of control, misunderstanding, unforeseen consequences, misuse, and abuse. Still, this strong "playing God" argument against AI urges us to think carefully about the ways in which the development and deployment of AI technologies could go seriously morally wrong. So, until we can make sure that we are able to control AI technologies, understand them and their effects, and use them benevolently, we have a strong case against developing and deploying AI technologies.

As we have seen in Chapter 3, our answers to the Critical Questions, CQ1–CQ3, will tell us what issues need to be addressed before the new or emerging technology in question can be developed, deployed, or used responsibly, that is, in accordance with our human limitations and without pretending that we are all-powerful, all-knowing, and all-good. In general, an affirmative answer to CQ1 suggests that the new or emerging technology in question could get out of the control of (less than omnipotent) human beings. Of course, we could never have perfect control over a new or emerging technology because we are not all-powerful as God is supposed to be. However, we can try to enhance our control over a new or emerging technology by analyzing risks and designing for safety. Accordingly, a "yes" to CQ1 suggests that we should direct our efforts toward risk analysis and safety. That is, we should conduct extensive risk analyses and safety testing, as well as design safety mechanisms informed by the results of these risk assessments and safety tests, to make sure that the new or emerging technology in question would be controllable by us before it can be deployed or used responsibly, that is, in accordance with our human limitations and without pretending that we are all-powerful.

An affirmative answer to CQ2 suggests that the new or emerging technology in question could be misunderstood or have unforeseen consequences for (less than omniscient) human beings. Of course, we could never have perfect understanding of a new or emerging technology and know all of its potential impacts because we are not all-knowing as God is supposed to be. However, we can try to improve our knowledge and understanding of a new or emerging technology through further research. Accordingly, a "yes" to CQ2 suggests that we should direct our efforts toward research. That is, we should conduct extensive experiments and simulations to learn as much as we can about the new or emerging technology in question before it can be deployed or used responsibly, that is, in accordance with our human limitations and without pretending that we are all-knowing.

An affirmative answer to CQ3 suggests that the new or emerging technology in question could be misused or abused by (less than omnibenevolent) human beings. Of course, we could never deploy or use a new or emerging technology in perfectly good ways because we are not all-good as God is supposed to be. However, we can try to prevent misuse and abuse by enacting regulations, policies, and laws with robust enforcement mechanisms. Accordingly, a "yes" to CQ3 suggests that we should direct our efforts toward policy. That is, we should conduct extensive deliberations, which would include all stakeholders, aimed at drafting regulations or legislating laws pertaining to the new or emerging technology in question before it can be deployed or used responsibly, that is, in accordance with our human limitations and without pretending that we are all-good.

Now, as far as the "playing God" argument against AI is concerned, we have answered Critical Questions, CQ1–CQ3, in the affirmative. This gives us good reasons to be concerned about AI (not only AGI or superintelligence but also strong/broad AI and even weak/narrow AI) getting out of control, being misunderstood or having unintended, negative consequences, and being misused or abused. Having good reasons to believe this would, in turn, help us to direct our governance efforts as far as AI is concerned. First, our affirmative answer to CQ1 suggests that we should direct our governance efforts toward conducting risk assessments and safety tests, as well as designing safety mechanisms, including cybersecurity measures, informed by the results of these risk assessments and safety tests, for AI technologies. Second, our affirmative answer to CQ2 suggests that we should direct our governance efforts toward further research on AI technologies and their potentially harmful impacts on us, our environment, and our planet. Third, our affirmative answer to CQ3 suggests that we should direct our governance efforts toward crafting policies for the development and deployment of AI technologies, so as to prevent misuse and abuse by cybercriminals, hackers, and other bad actors. In that respect, the "playing God" argument against AI, along with our answers to the Critical Questions, CQ1–CQ3, may provide some general guidance to AI governance.[8]

To sum up, here is how our answers to Critical Questions, CQ1–CQ3, can provide some general guidance for approaching the governance of AI.

CQ1: Could AI get out of the control of (less than omnipotent) human beings?

- Yes: attempt to enhance control by focusing on safety through risk assessment, safety testing, etc.

CQ2: Could AI be misunderstood or have unforeseen consequences for (less than omniscient) human beings?

- Yes: attempt to increase knowledge and understanding of consequences by focusing on research through experimentation, simulation, etc.

CQ3: Could AI be misused or abused by (less than omnibenevolent) human beings?

- Yes: attempt to prevent misuse or abuse by focusing on policy through regulation, legislation, enforcement, etc.

Notes

1 Relatedly, AI technologies drive the social media algorithms that tend to spread misinformation, disinformation, and fake news at the expense of accurate information on social media platforms. See Chapter 3.

2 As an anonymous reviewer pointed out, even if we are able to explain what a DNN is doing "under the hood," so to speak, it may be tackling a complex problem or processing vast amounts of data that are just too complicated for us to grasp.

3 I say more about Lethal Autonomous Weapon Systems (LAWS) in Chapter 8.

4 Related to AI-generated misinformation is the misinformation that tech companies themselves create and spread about the capabilities of their AI technologies. This is commonly known as "AI Hype," that is, "the overestimation and misrepresentation of AI capabilities" (Duarte et al. 2024, 649). According to Duarte et al. (2024, 649), "the negative aspect of AI Hype [is] associated with the staunch overestimation, exaggeration, and plain misrepresentation of AI found in news stories, press releases, and interviews that go beyond excitement into a realm of fantasy, manipulation, and speculation as fact." For example, when Sam Altman, CEO of OpenAI, boldly declares, without any evidence, that "95% of what marketers use agencies, strategists, and creative professionals for today will easily, nearly instantly and at almost no cost be handled by the AI—and the AI will likely be able to test the creative against real or synthetic customer focus groups for predicting results and optimizing" (Nicastro 2024), he is engaging in AI hype. Such AI hype may lead businesses to rush to adopt GenAI applications, which could expose them to greater cybersecurity risks (Humphreys et al. 2024).

5 I say more about AI and autonomous vehicles in Chapter 7.

6 The AI Risk Repository is free to copy and use: https://airisk.mit.edu/.

7 I say more about Lethal Autonomous Weapon Systems (LAWS) in Chapter 8.

8 Recall that the answers to the Critical Questions, CQ1–CQ3, of a "playing God" argument with respect to a new or emerging technology are meant to provide general guidance for approaching the governance of that technology. They are not supposed to constitute concrete governance recommendations or serve as a comprehensive governance proposal. On concrete governance proposals for the governance of AI, such as AI4People (Floridi 2023) and the OECD AI Principles (OECD 2024), see Huw Roberts et al. (2024). See also European Parliament (2024) on the EU AI Act.

7

Are We "Playing God" with Autonomous Vehicles?

In San Francisco, a Cruise self-driving car drove over a pedestrian who was hit by another, nonautonomous car (Ingram 2023). The pedestrian was dragged underneath the autonomous vehicle, which then stopped with her body underneath its rear axle, just before emergency personnel arrived on the scene. This incident is only one of many since Cruise, which is a subsidiary of General Motors, and Waymo, which is a subsidiary of Alphabet, Inc., started to operate fleets of robotaxis, which are autonomous vehicles with no human drivers, in San Francisco. After the incident, Cruise issued a statement that read, "The initial impact was severe and launched the pedestrian directly in front of the AV [i.e., Autonomous Vehicle]. *The AV then braked aggressively to minimize the impact*" (emphasis added).[1]

As we have seen in Chapter 6, outsourcing or delegating decision-making, including making decisions on the road, such as whether to brake or to accelerate, and so on, to technology raises the specter of "playing God." Indeed, in an interview for the *Daily Express*, the CEO of Regtransfers, Mark Trimbee, said the following about autonomous vehicles (Chillingsworth 2023):

> Who dictates the logic behind the vehicle's choices when faced with dilemmas that may harm either pedestrians or passengers? *Manufacturers of self-driving car intelligence and their decision-making algorithms could find themselves in the position of "playing God"*—and that's a problem (emphasis added).

Along the same lines, Mark Salisbury (2023) writes:

> If a self-driving car is left with two choices, both of which lead to the injury of either pedestrians or the passenger, *who decides the logic of its choices,*

> and why? Where such instances occur, the manufacturers of the intelligence behind self-driven cars—and the logic used—could be seen as "*playing God*" (emphasis added).[2]

These are instances of "playing God" arguments about self-driving, driverless, or autonomous vehicle technology. For many, "the idea of an autonomous car [. . .] playing God is disturbing" (Yoshida 2015).

So, let's consider carefully the "playing God" argument against autonomous vehicles. Using my argumentation scheme for "playing God" arguments (see Chapter 3), such a "playing God" argument against autonomous vehicles can be stated as follows.

> *Major Premise*: We should not (morally speaking) "play God."
> *Minor Premise*: Developing and deploying autonomous vehicles amount to "playing God."
> *Conclusion*: We should not (morally speaking) develop and deploy autonomous vehicles.

Before the conclusion of this argument can be accepted or rejected, we should evaluate this argument by asking the Critical Questions that accompany the argumentation scheme for "playing God" arguments. These questions are as follows.

> CQ1: Could autonomous vehicles get out of the control of (less than omnipotent) human beings?
> CQ2: Could autonomous vehicles be misunderstood or have unforeseen consequences for (less than omniscient) human beings?
> CQ3: Could autonomous vehicles be misused or abused by (less than omnibenevolent) human beings?

If the answer to all of these Critical Questions is "yes," then it means that developing and deploying autonomous vehicles would indeed amount to "playing God" in terms of power, knowledge, and benevolence. Given that we should not pretend to be all-powerful, all-knowing, and all-good as God is supposed to be, we could then conclude that we should not develop and deploy autonomous vehicles. In that case, the "playing God" argument against autonomous vehicles would be a strong argument. On the other hand, if the answer to all of these Critical Questions is "no," then it means that developing and deploying autonomous vehicles would not amount to "playing God." In that case, the conclusion that we should not develop and deploy autonomous vehicles would not be warranted and the "playing God" argument against autonomous vehicles would be a weak argument. Any other combination of "yes" and "no" answers to the Critical Questions, CQ1–CQ3, would make the "playing God" argument against autonomous vehicles stronger or weaker. That is, the more Critical Questions are answered in the affirmative, the stronger the argument, whereas the more Critical Questions are answered in the negative, the weaker the argument (see Table 3.1 in Chapter 3).

First, let's consider CQ1. Could autonomous vehicles get out of the control of (less than omnipotent) human beings? There are good reasons to answer CQ1 in the affirmative. Since the technologies that make vehicles self-driving or autonomous are AI technologies, for the most part, many of the issues with AI technologies discussed in Chapter 6 carry over to autonomous vehicles as well, including the problem of control. As Mary Cummings (2023), Professor of Systems Engineering who was the senior safety advisor to the National Highway Traffic Safety Administration, puts it:

> *[T]he AI that runs vehicles*—both the cars that operate in actual self-driving modes and the much larger number of cars offering advanced driving assistance systems (ADAS)—are *based on the same principles as ChatGPT and other large language models* (LLMs). These systems control a car's lateral and longitudinal position—to change lanes, brake, and accelerate—without waiting for orders to come from the person sitting behind the wheel (emphasis added).

For instance, as we have seen in Chapter 6, AI technologies could gradually rob us of our power and control "by stealthily taking away decision-making power app by app, or by diminishing our capacities as we rely more and more on machines to do our thinking for us" (Boddington 2023, 26–7). That is, as autonomous systems, such as self-driving cars, increasingly make more of our decisions for us, we would gradually lose control over making such decisions. On the road, human drivers have to make numerous decisions, such as whether to slow down or speed up in order to avoid a collision, and so on. If we let autonomous systems like self-driving cars make such decisions for us, we surrender our decision-making power and control over such decisions to these technologies. After several generations of outsourcing driving to self-driving cars, we could end up in a situation in which humans would lack the ability to operate vehicles even if they needed to do so due to some technological calamity.

Another thing we could lose control over as autonomous vehicles become more prevalent is our responsibility. That is, if a car is driverless, either because it is empty of humans or because all the humans inside it are designated as passengers (as in the case of Cruise and Waymo robotaxis), then it is unclear who bears responsibility (be it moral or legal responsibility) when something goes wrong. Typically, on the road, responsibility comes from being the driver of a vehicle who exercises control over that vehicle. But with self-driving cars, there is no driver, which makes it unclear who is responsible for driving safely. As Hansson, Belin, and Lundgren (2021, 1385) put it:

> *If there is no driver who controls the vehicle, who is then responsible for the safety of its passengers and of those who travel or walk on the same roads? If a car is "driven" by a computer possessing artificial intelligence, does that intelligence constitute an entity that can be held responsible?* What are the responsibilities of the vehicle's current owner? Its manufacturer? The owner and manager of the road system? The organization running the traffic control centre that the vehicle communicates with? (emphasis added)

As passengers in a self-driving car, we have no control over the autonomous vehicle, and thus no control over the operation of that vehicle, including braking, accelerating, turning, changing lanes, and so on. This loss of control robs us not only of our decision-making power but also of our responsibility in case of an accident.

From all of the above, it is reasonable to conclude that autonomous vehicles could get out of the control of (less than omnipotent) beings like us. By surrendering our control over the operation of a vehicle to an autonomous system, we could lose our power to make decisions on the road and be held responsible (be it morally or legally responsible) for the consequences of those decisions. Accordingly, the answer to CQ1, as far as autonomous vehicles are concerned, seems to be "yes" overall.

Next, let's consider CQ2. Could autonomous vehicles be misunderstood or have unforeseen consequences for (less than omniscient) human beings? There are also good reasons to answer CQ2 in the affirmative. The so-called "black box problem" in AI is a problem as far as autonomous vehicles are concerned as well. As we have seen in Chapter 6, some AI systems are black boxes, which means that we understand what goes into these systems (i.e., the input), and we understand what comes out of these systems (i.e., the output), but we do not understand what happens inside these systems. In other words, the internal workings of black boxes are opaque to us. Now, autonomous driving systems are often based on machine learning algorithms and Deep Neural Networks (DNNs), which are black boxes to us. For instance, computer vision is the branch of AI that deals with training neural networks to identify objects in images and videos. Such deep neural networks are trained to extract patterns from visual data and can be used to operate self-driving cars. This implies that, for us, the reason why a self-driving car brakes rather than accelerates, turns right rather than left, and so on, remains largely mysterious. We understand the input, that is, the visual data used to train the system, and we understand the output, for example, the self-driving car braking or turning right, but we do not understand how the system processed the data to arrive at the decision to brake or turn right. Consequently, a self-driving car could behave differently on the same stretch of road at different times of the day and we would have no idea why.

When it comes to autonomous vehicles, one poorly understood phenomenon is so-called "phantom braking." This is when a self-driving car suddenly brakes hard for no apparent reason. As Cummings (2023) puts it:

> *One failure mode not previously anticipated is phantom braking.* For no obvious reason, a self-driving car will suddenly brake hard, perhaps causing a rear-end collision with the vehicle just behind it and other vehicles further back. Phantom braking has been seen in the self-driving cars of many different manufacturers and in ADAS-equipped cars as well. *The cause of such events is still a mystery* (emphasis added).

The autonomous vehicles of all tech companies that utilize computer vision and other AI systems are prone to this problem. According to Cummings

(2023), autonomous vehicles could fail in countless other ways that we would be unable to foresee.

Moreover, some autonomous vehicles require wireless connectivity to operate properly. When connectivity drops, the consequences could be unforeseen and potentially harmful. For example, a driver who stopped behind two Cruise self-driving cars at an intersection got trapped when the Cruise self-driving cars in front of him would not move and more Cruise self-driving cars pulled up behind him. This driver, another driver, and "a paratransit bus were trapped in a robotaxi sandwich" (Marshall 2022). This incident may have been a case of connectivity loss, but we cannot say that with high confidence because the companies that own and operate robotaxis do not make such information publicly available. An anonymous letter from a Cruise employee to *Wired* alleges that "the company loses contact with its driverless vehicles with regularity, blocking traffic and potentially hindering emergency vehicles" (Marshall 2022).

Autonomous vehicles also carry pretty much the same uncertain environmental risks that come with AI technologies in general. As we have seen in Chapter 6, we know that AI has a massive carbon footprint, but we do not know how massive it is (Floridi 2023, 183). The data centers that store the data on which AI systems are trained, including the AI systems used in autonomous vehicles, consume vast amounts of energy and potable water. The environmental impacts of such increasing energy and water consumption are uncertain.

In addition to uncertain environmental impacts, the social and economic impacts that autonomous vehicles would have are also uncertain. For instance, autonomous vehicles could take jobs away from human drivers. Autonomous vehicles could also make parking for nonautonomous vehicles less abundant as tech companies expand their fleets of autonomous vehicles in more cities. For instance, Tesla is also getting into the robotaxi business as the company recently unveiled its "cybercab" (Hawkins 2024).[3] Likewise, owners of autonomous vehicles could try to avoid paying for parking by having their autonomous vehicles cruise the streets, thereby exacerbating congestion on city roads (Choi 2019). Autonomous vehicles could have many other unforeseen impacts and failure modes that we simply have not even conceived of yet.

From all of the above, it is reasonable to conclude that autonomous vehicles could be misunderstood or have unforeseen consequences for (less than omniscient) human beings. Many of the AI systems that make vehicles autonomous are black box systems, which is why the black box problem in AI is a problem for autonomous vehicles as well. This means that autonomous vehicles could act in ways that we fail to understand or foresee (e.g., phantom braking). In addition, connectivity issues, as well as the energy and water needed to train the AI systems used to make vehicles autonomous or cars self-driving, could lead to unforeseen consequences. The social and economic consequences of autonomous vehicles could also be unforeseen for, and poorly understood by, less than all-knowing beings like us. They could cost human drivers their jobs, parking spaces, and more. Accordingly, the answer to CQ2, as far as autonomous vehicles are concerned, seems to be "yes" overall.

Let's move on to CQ3. Could autonomous vehicles be misused or abused by (less than omnibenevolent) human beings? There are several ways in which autonomous vehicles could be misused or abused by less than all-good beings like us. This is because autonomous vehicles are dual use technologies insofar as they can be (mis)used (or abused) in ways other than their intended or designed use (see Chapter 3). In fact, autonomous vehicles have already been used to commit crimes. For example, Unmanned Underwater Vehicles (UUVs) have been used for drug trafficking (Floridi 2023, 124). In July 2022, Spanish law enforcement seized autonomous UUVs that are "capable of carrying up to 200kg (441lbs) of cargo" (Sands 2022). Equipped with GPS systems, these UUVs "could potentially be operated by drug traffickers from anywhere in the world using an internet-device as simple as a tablet" (Sands 2022).

Like other AI systems, autonomous vehicles are powered by machine learning algorithms that require vast amounts of data in order to function properly. In the course of their operation, autonomous vehicles continue to collect data through their cameras, sensors, microphones, and so on, which effectively makes autonomous vehicles rather lucrative targets for hackers, cybercriminals, and other bad actors. As the Director of the Federal Bureau of Investigation (FBI), Christopher Wray, put it when discussing the security risks posed by autonomous vehicles, "any time you aggregate lots and lots of sensitive data, it makes a very tempting target" (C-SPAN 2023). Along similar lines, Hansson, Belin, and Lundgren (2021, 1403) observe:

> Extensive information about routes and destinations will have to be collected in order to optimize the movements of self-driving vehicles. *Such information can be misused or hacked. It can for instance be used to convey commercial and political messages to car users. An authoritarian state can use it to keep track of the opposition* (emphasis added).

As aggregators of large amounts of data, autonomous vehicles could also be used as surveillance platforms by state governments to spy on citizens, by employers to spy on their employees, and so on.

Once hacked, autonomous vehicles could be used to create all sorts of havoc on the road, including traffic jams, automobile accidents, and the like. Consider, for example, bank robbers using autonomous vehicles to block major intersections on the road to the bank they are robbing, thereby making it difficult for law enforcement to get to the crime scene on time before the robbers make their escape. Or consider, for example, human traffickers using autonomous vehicles, which are disguised as robotaxis, to kidnap people. Autonomous vehicles could be not only criminalized (i.e., turned into vehicles of crime) but also weaponized (i.e., turned into vehicles of harm). For instance, the Director of the FBI, Christopher Wray, described autonomous vehicles "as a potentially new 'attack' surface for terrorists to use to harm civilians" (Straughan 2023).[4]

Another way in which autonomous vehicles could have a corrupting influence on us is what Keith Abney (2022) calls the "chicken problem." That is, human drivers could try to "play chicken" with autonomous vehicles. Human drivers might believe that a self-driving car's prime directive is to avoid collisions, which might lead a human driver to believe that a self-driving car would swerve at the last minute to avoid a collision. In that case, a human driver could steer directly at a self-driving car on purpose, thereby putting human lives and property in harm's way. Similarly, human pedestrians might believe that an autonomous vehicle's prime directive is to avoid hitting human pedestrians, which might lead a human pedestrian to stand in the way of, or even walk toward, an autonomous vehicle instead of moving out of its way. Apparently, this has already happened (Hawkins 2018).

> Last night, Greylock partner Josh Elman had an encounter with one of Waymo's self-driving cars outside his California home. He saw the minivan approaching as he was taking his trash to the curb. But rather than stand passively to the side as the vehicle passed like you would do with literally any other car, *Elman says he ran toward it* (emphasis added).

As Abney (2022, 260) states the problem, "*how should an AV deal with abusive behavior by other drivers, including playing chicken with the car?*" (original emphasis). Although it might seem obvious, Andrew Hawkins (2018) pleads with people to not play chicken with autonomous vehicles.

> *Please don't play chicken with the self-driving cars.* These are 4,000-pound steel machines. You are a sack of meat and sticks. It's not a fair match (emphasis added).

This is an example of how emerging technologies, such as autonomous vehicles, could change our behavior for the worse, more specifically, by resulting in people engaging in abusive behavior that could put their own lives as well as the lives of others at risk.

Autonomous vehicles could also lead the tech companies that make them to behave badly as well. Missy Cummings, formerly the senior safety advisor to the National Highway Traffic Safety Administration, recounts that she has witnessed faulty reporting on accidents involving self-driving cars from the companies that make them. When she watched video footage of accidents involving self-driving cars, the videos did not always match the reports of the companies whose self-driving cars were involved. In particular, the companies' reports tended to put the blame on human drivers rather than on the self-driving cars (Hsu 2024). Accident reports showing bias against human drivers and in favor of self-driving cars could paint a misleading picture that self-driving cars are safer than human drivers. Such a misleading picture could, in turn, support a false belief that the technology is ready for deployment at scale, thereby leading to more self-driving cars being deployed on city streets. This would be a misuse of autonomous vehicles if the state of this technology does not yet warrant full-scale deployment.

From all of the above, it is reasonable to conclude that autonomous vehicles could be misused or abused by (less than omnibenevolent) human beings. Autonomous vehicles pose a security risk because they could be hacked and controlled remotely by bad actors. They could also be turned into vehicles of crime (e.g., drug trafficking, kidnapping, etc.), vehicles of harm (e.g., carrying out terrorist attacks, playing chicken, etc.), and surveillance platforms. Accordingly, the answer to CQ3, as far as autonomous vehicles are concerned, seems to be "yes" overall.

As far as autonomous vehicles are concerned, Critical Questions, CQ1–CQ3, seem to have affirmative answers (see Table 7.1 for a summary). This means that developing and deploying autonomous vehicles would indeed amount to "playing God" in terms of power (CQ1), knowledge (CQ2), and benevolence (CQ3). Therefore, a "playing God" argument against autonomous

Table 7.1 Summary of the Answers (with Justifications) to the Critical Questions, CQ1–CQ3, of the "Playing God" Argument against Autonomous Vehicles

Critical Question	Answer	Justification
CQ1: Could autonomous vehicles get out of the control of (less than omnipotent) human beings?	yes	the control problem (AI); loss of power and control through the outsourcing of decision-making to autonomous vehicles; loss of control over responsibility for accidents involving autonomous vehicles
CQ2: Could autonomous vehicles be misunderstood or have unforeseen consequences for (less than omniscient) human beings?	yes	the black box problem (AI); unexplained and unforeseen failures of vehicle operation; unpredictable behavior of autonomous vehicles (e.g., phantom braking); uncertain environmental impacts from increasing energy and water consumption to train autonomous vehicles; uncertain social and economic impacts (e.g., loss of jobs, parking spaces, etc.)
CQ3: Could autonomous vehicles be misused or abused by (less than omnibenevolent) human beings?	yes	autonomous vehicles could be criminalized (e.g., drug trafficking); autonomous vehicles could be hacked (e.g., privacy violations); autonomous vehicles could be weaponized (e.g., terrorism); the chicken problem

Source: Author

vehicles is a strong argument (see Table 3.1 in Chapter 3). This argument can be summed up as follows.

Major Premise: We should not (morally speaking) "play God."
Conditional Premise: If autonomous vehicles could get out of the control of humans, be misunderstood or have unforeseen consequences for humans, or be misused or abused by humans, then developing and deploying autonomous vehicles amount to "playing God."
Control Premise: Autonomous vehicles could get out of the control of humans.
Understanding Premise: Autonomous vehicles could be misunderstood or have unforeseen consequences for humans.
Use Premise: Autonomous vehicles could be misused or abused by humans.
Conclusion: We should not (morally speaking) develop and deploy autonomous vehicles.

This strong "playing God" argument against autonomous vehicles points to several moral problems with these technologies, namely, that they could get out of control, that the way they work could be misunderstood, that they could have unintended long-term consequences for us and the environment, and that they could be misused or abused by people.

Like all "playing God" arguments, this one against autonomous vehicles is a defeasible argument. As we have seen in Chapter 3, conclusions of defeasible arguments can always be defeated by additional evidence that is not cited in their premises. In other words, this argument is not the final word about whether or how we should develop and deploy autonomous vehicles. One obvious way to defeat this "playing God" argument against autonomous vehicles is to argue that the potential benefits of these technologies would outweigh the perils of loss of control, misunderstanding, unforeseen consequences, misuse, and abuse. Still, this strong "playing God" argument against autonomous vehicles urges us to think carefully about the ways in which the development and deployment of autonomous vehicles could go seriously morally wrong. So, until we can make sure that we are able to control these technologies, understand them and their effects, and use them benevolently, we have a strong case against developing and deploying autonomous vehicles.

As we have seen in Chapter 3, our answers to the Critical Questions, CQ1–CQ3, will tell us what needs to be addressed before the new or emerging technology in question can be developed, deployed, or used responsibly, that is, in accordance with our human limitations and without pretending that we are all-powerful, all-knowing, and all-good. In general, an affirmative answer to CQ1 suggests that the new or emerging technology in question could get out of the control of (less than omnipotent) human beings. Of course, we could never have perfect control over a new or emerging technology because we are not all-powerful as God is supposed to be. However, we can try to enhance our control over a new or emerging technology by analyzing risks and designing for safety. Accordingly, a "yes" to CQ1 suggests that we should direct our efforts toward risk analysis and safety. That is, we should conduct extensive risk analyses and safety testing, as well as design safety mechanisms informed by the results of these risk assessments and safety tests, to make sure that the new or emerging technology in question would be controllable by us before it can be deployed or used responsibly, that is, in accordance with our human limitations and without pretending that we are all-powerful.

An affirmative answer to CQ2 suggests that the new or emerging technology in question could be misunderstood or have unforeseen consequences for (less than omniscient) human beings. Of course, we could never have perfect understanding of a new or emerging technology and know all of its potential impacts because we are not all-knowing as God is supposed to be. However, we can try to improve our knowledge and understanding of a new or emerging technology through further research. Accordingly, a "yes" to CQ2 suggests that we should direct our efforts toward research. That is, we should conduct extensive experiments and simulations to learn as much as we can about the new or emerging technology in question before it can be deployed or used responsibly, that is, in accordance with our human limitations and without pretending that we are all-knowing.

An affirmative answer to CQ3 suggests that the new or emerging technology in question could be misused or abused by (less than omnibenevolent) human beings. Of course, we could never deploy or use a new or emerging technology in perfectly good ways because we are not all-good as God is supposed to be. However, we can try to prevent misuse and abuse by enacting regulations, policies, and laws with robust enforcement mechanisms. Accordingly, a "yes" to CQ3 suggests that we should direct our efforts toward policy. That is, we should conduct extensive deliberations, which would include all stakeholders, aimed at drafting regulations or legislating laws pertaining to the new or emerging technology in question before it can be deployed or used responsibly, that is, in accordance with our human limitations and without pretending that we are all-good.

Now, as far as the "playing God" argument against autonomous vehicles is concerned, we have answered Critical Questions, CQ1–CQ3, in the affirmative. This gives us good reasons to be concerned about autonomous vehicles getting out of control, being misunderstood or having unintended, negative consequences, and being misused or abused. Having good reasons to believe this would, in turn, help us to direct our governance efforts as far as autonomous vehicles are concerned. First, our affirmative answer to CQ1 suggests that we should direct our governance efforts toward conducting risk assessments and safety tests, as well as designing safety mechanisms, including cybersecurity measures, informed by the results of these risk assessments and safety tests, for autonomous vehicles. Second, our affirmative answer to CQ2 suggests that we should direct our governance efforts toward further research on the perils of autonomous vehicles for our societies, economies, and environments. Third, our affirmative answer to CQ3 suggests that we should direct our governance efforts toward drafting rules of the road for autonomous vehicles, so as to prevent misuse and abuse by hackers, cybercriminals, terrorists, and other bad actors. In that respect, the "playing God" argument against autonomous vehicles, along with our answers to the Critical Questions, CQ1–CQ3, may provide some general guidance to autonomous vehicles governance.[5]

To sum up, here is how our answers to Critical Questions, CQ1–CQ3, can provide some general guidance for approaching the governance of autonomous vehicles.

CQ1: Could autonomous vehicles get out of the control of (less than omnipotent) human beings?

- Yes: attempt to enhance control by focusing on safety through risk assessment, safety testing, etc.

CQ2: Could autonomous vehicles be misunderstood or have unforeseen consequences for (less than omniscient) human beings?

- Yes: attempt to increase knowledge and understanding of consequences by focusing on research through experimentation, simulation, etc.

CQ3: Could autonomous vehicles be misused or abused by (less than omnibenevolent) human beings?

- Yes: attempt to prevent misuse or abuse by focusing on policy through regulation, legislation, enforcement, etc.

Notes

1 Since 2016, "automotive AI in the United States has been linked to at least 25 confirmed deaths and to hundreds of injuries and instances of property damage" (Cummings 2023).

2 Cf. Luvuyo Gantsho (2022).

3 Concerns have already been expressed with respect to Tesla's cybercab. Tesla's cybercab uses so-called "end-to-end" AI technology, which involves training an autonomous system to make decisions from raw data directly, that is, with no intermediate steps that require additional engineering or programming. This is a safety concern, as one Tesla engineer said, because "end-to-end AI technology is a black box, making it nearly impossible to see what went wrong when it misbehaves and causes an accident" (Shirouzu and Kirkham 2024). On the black box problem in AI, see Chapter 6.

4 I say more about Lethal Autonomous Weapon Systems (LAWS) in Chapter 8.

5 Recall that the answers to the Critical Questions, CQ1–CQ3, of a "playing God" argument with respect to a new or emerging technology are meant to provide general guidance for approaching the governance of that technology. They are not supposed to constitute concrete governance recommendations or serve as a comprehensive governance proposal. On concrete governance proposals for the governance of autonomous vehicles, see Araz Taeihagh and Hazel Si Min Lim (2019).

8

Are We "Playing God" with Lethal Autonomous Weapon Systems?

HARPY, or Anti-Radiation Loitering Munition, is a fully autonomous weapon system developed by the Israel Aerospace Industries (IAI). According to the IAI (2023), it is the "world's first and most operational Anti-Radiation Loitering Munition system." Also known as "suicide drones," "kamikaze drones," or "exploding drones," loitering munitions are a type of Unmanned Combat Aerial Vehicle (UCAV) equipped with explosive warheads. They can loiter around an area, select a target, and then crash into it, thereby destroying the target. According to the IAI (2023), the "HARPY can *autonomously seek and strike* emitting, high-value targets. [. . .] HARPY is *equipped to hunt—seek targets in a designated area*, locate and identify their frequency, and *autonomously pursue a strike* from any direction, at shallow or steep dive profiles" (emphasis added).

Like the HARPY, the IAI's HAROP, or Long Range Loitering Munition, is also an autonomous weapon system. Instead of an anti-radiation sensor, the HAROP has an electro-optic sensor, which enables it to seek and destroy more targets than just air defenses. More importantly, while the HARPY is a so-called "fire and forget" autonomous weapon, the HAROP is supervised by remote human-in-the-loop mission control and thus can be aborted by a human operator. The HAROP was used by Azerbaijan as it waged war in the Armenian region of Nagorno-Karabakh in 2020. "Azerbaijan won the 2020 war in a little over a month, regaining much of the territory internationally recognized as part of Azerbaijan but populated and governed, until now, almost exclusively by ethnic Armenians, following the expulsion of ethnic Azeris in the late 1980s and early 1990s" (Saltman 2023). Azerbaijani officials said that autonomous weapons like the HAROP were "integral to their country's success in Nagorno-Karabakh during the 2020 war" (Saltman 2023).

Using AI to make weapon systems autonomous is one of the ways in which AI technologies can deviate from their intended or designed use. As we have seen in Chapter 6, AI technologies are dual use technologies insofar as they can be (mis)used (or abused) in ways other than their intended or designed use. As dual use technologies, then, AI systems can be weaponized. Autonomous weapons can vary in their level of autonomy, that is, their ability to execute a task or set of operations without human intervention, from semiautonomous to fully autonomous, either with or without human supervision. In that respect, Human Rights Watch (2012) provides the following useful classifications:

- Human-*in*-the-Loop Weapons: Robots that can select targets and deliver force only with a human command;
- Human-*on*-the-Loop Weapons: Robots that can select targets and deliver force under the oversight of a human operator who can override the robot's actions;
- Human-*out-of*-the-Loop Weapons: Robots that are capable of selecting targets and delivering force without any human input or interaction.

Autonomous weapons can also vary by the type of decision or function that is autonomous insofar as it is not controlled by human operators. In US Department of Defense Directive 3000.09, the Office of the Under Secretary of Defense for Policy (2023, 21) defines an autonomous weapon system as "A weapon system that, once activated, *can select and engage targets without further intervention by an operator*. This includes, but is not limited to, operator supervised autonomous weapon systems that are designed to allow operators to override operation of the weapon system, but can select and engage targets without further operator input after activation" (emphasis added).

Lethal Autonomous Weapon Systems (LAWS) "are lethal devices that identify potential enemy targets and independently choose to attack those targets on the basis of algorithms and AI" (Dresp-Langley 2023, 3). LAWS are distinguished from other types of autonomous weapons that are used mostly for defensive purposes, such as air defense systems like the Iron Dome, rather than offensive purposes. Amoroso and Tamburrini (2020, 188) provide the following useful examples:

> (i) *air defense systems* (e.g., Phalanx, MANTIS, Iron Dome, Goalkeeper); (ii) *active protection systems*, which shield armored vehicles by identifying and intercepting anti-tank missiles and rockets (e.g., LEDS-150 and Trophy); (iii) *robotic sentries*, like the Super aEgis II stationary robotic platform tasked with the surveillance of the demilitarized zone between North and South Korea; (iv) *guided munitions*, which autonomously identify and engage targets that are not in sight of the attacking aircraft (e.g., the Dual-Mode Brimstone); and (v) *loitering munitions*, such as the Harpy NG, which overfly an assigned area in search of targets to dive-bomb and destroy (original emphasis).

In this chapter, I focus on Lethal Autonomous Weapon Systems (LAWS), such as loitering munitions like the IAI's HARPY, which are capable of making and executing targeting decisions without human intervention.

In addition to lethal autonomous drones, such as the HARPY, other prominent examples of LAWS include lethal autonomous drone swarms as well as so-called "killer robots" or "slaughter bots." As we have seen, lethal autonomous drones have already been developed and deployed in battle, as in the 2020 war in Nagorno-Karabakh. Lethal autonomous drone swarms have already been developed and deployed as well. In May 2021, "Israel used Elbit Systems' Legion-X drone swarm in combat against Hamas, with the drones searching for targets and relaying information to the Israel Defense Forces" (Kallenborn 2024). This is widely believed "to be the first time a drone swarm has been used in combat" (Hambling 2021).

So-called "killer robots" or "slaughter bots" are lethal autonomous robotic systems that are "able to select and attack targets without intervention by a human operator" (Dresp-Langley 2023, 2). The 2017 short film, *Slaughterbots*, illustrates in quite vivid detail what an attack by a swarm of lethal autonomous drones could look like.[1] It presents a dystopian scenario in which swarms of lethal autonomous drones are able to search, find, and assassinate human targets by means of AI systems. Equipped with facial recognition technology, these swarming killer drones can target specific individuals. The short film was made as part of a campaign to stop LAWS known as the Campaign to Stop Killer Robots, which was formed in 2013 "to pre-emptively ban lethal robot weapons that would be able to select and attack targets without any human intervention" (Campaign to Stop Killer Robots 2013). At the end of the short film, the computer scientist Stuart Russell says, "allowing machines to choose to kill humans will be devastating to our security and freedom."

In his "A Hippocratic Oath for Artificial Intelligence Practitioners," Oren Etzioni (2018) identifies the decision to kill humans, or matters of life and death, as especially morally problematic, and thus in need of careful handling, before it can be delegated to autonomous systems. As Etzioni (2018) puts it:

> Most especially must I tread with care in *matters of life and death*. If it is given me to save a life using AI, all thanks. But *it may also be within AI's power to take a life*; this awesome responsibility must be faced with great humbleness and awareness of my own frailty and the limitations of AI. Above all, *I must not play at God nor let my technology do so* (emphasis added).

Accordingly, the design, development, deployment, and use of LAWS raise moral concerns about the power, sophistication, and potential for misuse or abuse of these technologies. For one thing, since they will be autonomous, LAWS will have "the power and discretion to select targets and take lives without human involvement," as the Secretary-General of the United Nations, António Guterres, put it (UN News 2019). In that sense, there is a danger of LAWS "going rogue." This concern is distinct from the concern that LAWS could "fall into the wrong hands," that is, be hacked and then used for nefarious purposes. For Guterres, "machines with the power and discretion to take lives without human involvement are politically unacceptable, morally repugnant and should be prohibited by international law" (UN News 2019).

Those who are concerned about LAWS getting out of control, having unforeseen consequences, or being misused or abused by humans are effectively making a "playing God" argument against such weapons. Using my argumentation scheme for "playing God" arguments (see Chapter 3), such a "playing God" argument against LAWS can be stated as follows.

> *Major Premise*: We should not (morally speaking) "play God."
> *Minor Premise*: Developing and deploying LAWS amount to "playing God."
> *Conclusion*: We should not (morally speaking) develop and deploy LAWS.

Before the conclusion of this argument can be accepted or rejected, we should evaluate this argument by asking the Critical Questions that accompany the argumentation scheme for "playing God" arguments. These questions are as follows.

CQ1: Could LAWS get out of the control of (less than omnipotent) human beings?
CQ2: Could LAWS be misunderstood or have unforeseen consequences for (less than omniscient) human beings?
CQ3: Could LAWS be misused or abused by (less than omnibenevolent) human beings?

If the answer to all of these Critical Questions is "yes," then it means that developing and deploying LAWS would indeed amount to "playing God" in terms of power, knowledge, and benevolence. Given that we should not pretend to be all-powerful, all-knowing, and all-good as God is supposed to be, we could then conclude that we should not develop and deploy LAWS. In that case, the "playing God" argument against LAWS would be a strong argument. On the other hand, if the answer to all of these Critical Questions is "no," then it means that developing and deploying LAWS would not amount to "playing God." In that case, the conclusion that we should not develop and deploy LAWS would not be warranted and the "playing God" argument against LAWS would be a weak argument. Any other combination of "yes" and "no" answers to the Critical Questions, CQ1–CQ3, would make the "playing God" argument against LAWS stronger or weaker. That is, the more Critical Questions are answered in the affirmative, the stronger the argument, whereas the more Critical Questions are answered in the negative, the weaker the argument (see Table 3.1 in Chapter 3).

First, let's consider CQ1. Could LAWS get out of the control of (less than omnipotent) human beings? There are good reasons to answer CQ1 in the affirmative. Since the technologies that make drones, robots, and other types of weapons autonomous are AI technologies, for the most part, many of the moral issues with AI technologies discussed in Chapter 6 carry over to LAWS as well. In fact, according to Russell, weaponizing AI systems is pretty straightforward. As Russell puts it, "The technical capability for a system to find a human being and kill them is much easier than to develop a self-driving car. It's a graduate-student project" (quoted in Adam 2024, 521). As we have seen in Chapter 6, AI technologies could steadily rob us of our power and control "by stealthily taking away decision-making power app by app, or by diminishing our capacities as we rely more and more on machines to do our thinking for us" (Boddington 2023, 26–7). That is, as autonomous systems, such as LAWS, increasingly make more of our decisions for us, we would gradually lose control over making such decisions. During times of war, human soldiers have to make numerous decisions, including identifying enemy targets, distinguishing between enemy combatants and noncombatants, selecting weapons and ammunitions to use against enemy forces, engaging enemy soldiers, initiating attacks, and so on. If we let LAWS make such complex decisions for us, we surrender our power and control over matters of life and death to these technologies.

As in the case of other autonomous systems, such as autonomous vehicles, another thing we could lose control over as LAWS are developed and deployed on the battlefield is our accountability. That is, if a weapon system is autonomous, then it is unclear who bears responsibility (be it moral or legal responsibility) when something goes wrong. For example, LAWS could fail to distinguish between enemy soldiers and civilians, or mistake enemy combatants for noncombatants, which could result in innocent civilians getting hurt or killed. Who should be held accountable when such mistakes occur? After all, "you can't hold the weapon system itself accountable," according to Lucy Suchman, whereas to "hold the operator responsible for the actions of a system that was functioning autonomously" would be "unfair" (quoted in Adam 2024, 523). As Amoroso and Tamburrini (2020, 189) put it:

> Autonomy in weapons systems would *hinder responsibility ascriptions in case of wrongdoings*, by removing human operators from the decision-making process (emphasis added).[2]

LAWS could make many other catastrophic mistakes on the battlefield, including mistakes that could cost human lives, and it would be unclear who should be held responsible for those deadly mistakes. "In 2007, for example, the UK military hastily redesigned its autonomous Brimstone missile for use in Afghanistan when it was feared it might mistake a bus of school-children for a truckload of insurgents" (Adam 2024, 523). If such a catastrophic mistake were to occur, it would not be a "freak accident," but rather a situation where "responsibility gaps" may arise (Wood 2023, 4). Of course, the ultimate mistake would involve LAWS "going rogue," not in the sense of having some sort of malfunction or breakdown, but rather in the sense of turning against and then destroying their creators or operators, Terminator style.[3]

From all of the above, it is reasonable to conclude that LAWS could get out of the control of (less than omnipotent) beings like us. By deploying LAWS, we could lose our power to make decisions on the battlefield, where one of the most morally significant decisions that can be made is the decision to kill human beings. By deploying LAWS, we could also lose control over who would be held responsible (be it morally or legally responsible) when deadly mistakes involving LAWS would happen on the battlefield. The ultimate loss of control over LAWS would involve the fully autonomous weapons turning against their creators and operators. Accordingly, the answer to CQ1, as far as LAWS are concerned, seems to be "yes" overall.

Next, let's consider CQ2. Could LAWS be misunderstood or have unforeseen consequences for (less than omniscient) human beings? There are good reasons to answer CQ2 in the affirmative as well. For the so-called "black box problem" in AI is a problem as far as LAWS are concerned, too. As we have seen in Chapter 6, the internal workings of AI systems are often opaque to us. As far as Deep Neural Networks (DNNs) are concerned, for instance, we do not understand how the numerous artificial neurons (nodes) that make up such networks work together to produce outputs, decisions, predictions, and so on. Oftentimes, it is unclear what individual artificial neurons do on their own to generate outputs. Now, LAWS are often based on DNNs and machine learning algorithms that are black boxes for us. For instance, recall the 2017 short film *Slaughterbots*,[4] in which swarms of autonomous drones equipped with facial recognition technology are able to target individual human beings. Apparently, the US Air Force has developed autonomous drones with AI-powered facial recognition technology (Hambling 2023). Now, facial recognition systems typically use Convolutional Neural Networks (CNNs), which are a type of Deep (learning) Neural Networks (DNNs) well-suited for image classification tasks. Like DNNs, CNNs are black boxes, which means that we know what goes into these systems (i.e., the input, which is visual data, or more specifically, images of faces in the case of face recognition), and we know what comes out of these systems (i.e., the output, which is a match between a target face and a face in some database), but we do not know what happens in between. The internal workings, or decision-making processes, of DNNs and CNNs are opaque to us. For this reason, the outputs of facial recognition systems are not interpretable by humans. As far as LAWS that utilize facial recognition technology are concerned, then, we would be unable to explain why an autonomous drone with AI-powered facial recognition technology, such as those developed by the US Air Force (Hambling 2023), selected a specific target to hit or a particular human being to kill. If LAWS are either entirely or partly black box systems, as facial recognition systems are, then their decision-making processes would not be transparent to us.

Moreover, facial recognition technology is notoriously unreliable. This is because face recognition algorithms are biased. The US National Institute of Standards and Technology (NIST) tested roughly two hundred face recognition algorithms on two common face recognition tasks, namely, "one-to-one" matching, which involves matching a picture of a face to another picture of the same face in a database, and "one-to-many" searching, which involves determining whether a picture of a face has a match in a database. The datasets for this comprehensive study included "18.27 million images of 8.49 million people" (Hao 2019). The findings "call into question whether these [facial recognition] systems should continue to be so widely used" (Hao 2019). For the findings show that face recognition algorithms have a high false positive rate, that is, "they were more likely to find a match when there wasn't one" (Hao 2019). In particular, "Algorithms developed in the US were all consistently bad at matching Asian, African-American, and Native American faces" (Hao 2019). So, if facial recognition algorithms are incorporated into LAWS to make those systems "able to select and attack targets without intervention by a human operator" (Dresp-Langley 2023, 2), as the US Air Force seems to have done (Hambling 2023), it is likely that such autonomous weapons would exhibit similar racial biases. But the extent to which LAWS would be so biased remains uncertain.

Furthermore, insofar as they are black box systems, either in whole or in part, LAWS could behave differently under similar battle conditions at different times and we would have no idea why. Worse still, under unconventional battle conditions, namely, unconventional warfare or unstructured warfare scenarios, it is rather difficult to predict how LAWS would behave, especially when they encounter enemy LAWS. In that respect, deploying LAWS could lead to unforeseen escalations in armed conflicts, "an increase in the pace of war," and "unpredictable interactions" between LAWS deployed in warfare (Sharkey 2019, 77). As Mark Gubrud (2016) puts it:

> Autonomous weapons could malfunction, kill innocents, and nobody be held responsible. *Which kind of situation would occur most often*, and whether autonomous weapons would be more or less deadly than their prohibition, assuming everything else would be the same, *is endlessly debatable* (emphasis added).

Accordingly, LAWS could fail in countless ways that would be unforeseen by us. Like other autonomous systems, LAWS generally require wireless connectivity to operate properly. When connectivity drops, the consequences could be unpredictable and dangerous. What is certain is that, when connectivity is lost, there would be no way for human operators to exercise meaningful human control over LAWS.

LAWS also carry the uncertain environmental risks that come with AI technologies in general. As we have seen in Chapter 6, we know that AI has a massive carbon footprint, but we do not know how massive it is (Floridi 2023, 183). The data centers that store the data on which AI systems are trained, including the AI systems used in LAWS, such as facial recognition algorithms, consume vast amounts of energy and potable water. Currently, the environmental impacts of such increasing energy and water consumption are indeterminate.

In addition to uncertain environmental impacts, the development and deployment of LAWS could also lead to heretofore unforeseen geopolitical consequences. According to Amoroso and Tamburrini (2020, 189):

> Autonomy in weapons systems would *threaten in special ways international peace and stability*, by making wars easier to wage on account of reduced

> numbers of involved soldiers, by laying the conditions for unpredictable interactions between AWS [i.e., autonomous weapon systems] and their harmful outcomes, and by accelerating the pace of war beyond human reactive abilities (emphasis added).

The design, development, deployment, and use of LAWS could have long-term effects on international stability and security. Furthermore, when LAWS fight our wars for us, those wars could really turn out to be endless wars. For the use of LAWS "could reduce the risks of a 'body bag count', and as a consequence, remove a major disincentive for war" (Sharkey 2019, 77). By making wars thus easier to wage, LAWS could destabilize the entire world and put peace out of reach. Even without the lethal element, Amoroso and Tamburrini (2018) argue, autonomous weapons could still have a destabilizing effect on a global scale if they were to be used to destroy infrastructure. For instance, using LAWS to destroy infrastructure could affect global supply chains and trade in yet unforeseen ways.

From all of the above, it is reasonable to conclude that LAWS could be misunderstood or have unforeseen consequences for (less than omniscient) human beings. Many of the AI systems that make lethal weapons autonomous are black box systems, which is why the black box problem in AI is a problem as far as LAWS are concerned as well. This means that LAWS could act in ways that we fail to understand (e.g., bias in face recognition, unexplained target selection, etc.). In addition, connectivity issues, as well as the energy and water consumption from training the AI systems used in making lethal weapons autonomous, could lead to unforeseen consequences. The geopolitical consequences of LAWS could also be unpredicted, and poorly understood, by less than all-knowing beings like us (e.g., a global destabilizing effect, endless wars, etc.). Accordingly, the answer to CQ2, as far as LAWS are concerned, seems to be "yes" overall.

Now, let's move on to CQ3. Could LAWS be misused or abused by (less than omnibenevolent) human beings? There are a few ways in which LAWS could be misused or abused by less than all-good human beings. This is partly because responsibility gaps give rise to opportunities for misuse and abuse. That is, as we deploy LAWS, we could lose our grip on who is responsible for any harm caused by LAWS. Now, when it becomes difficult, perhaps even impossible, "to identify anyone who can be held responsible for harm caused by LAWS" (Oimann 2023, 2), less than morally perfect beings like us tend to think that we could "get away with murder," so to speak, which in turn makes us more likely to act in ways we would otherwise not dare to act. In other words, hiding behind LAWS could have an effect on us that is comparable to that of wearing the Ring of Gyges. In book II of the *Republic*, one of Plato's characters, namely, Glaucon, tells the story of Gyges, a shepherd who cannot resist the temptation of doing wrong after he is given a ring that makes him invisible when he wears it. With this invisibility ring, Gyges seduces the queen, conspires with her to murder the king of Lydia, and takes the king's place as ruler. Like Gyges' invisibility ring, Walter Isaacson (2016) argues, online anonymity allows people to say things online that they would never say to another person's face. Like Gyges' invisibility ring or the cloak of online anonymity, then, LAWS could also enable bad behavior by state actors that would exploit the responsibility gaps created by such weapons to evade accountability for any harm caused by LAWS.[5]

Relatedly, state actors could also use LAWS to inflict more harm than is necessary to achieve some military objective. According to Article 35.2 of the Additional Protocols to the Geneva Convention (Diplomatic Conference on the Reaffirmation and Development of International Humanitarian Law Applicable in Armed Conflicts 1977), "It is prohibited to employ weapons, projectiles and material and methods of warfare of a nature to cause *superfluous injury* or *unnecessary suffering*" (emphasis added). This principle of necessity dictates that any weapons that cause unnecessary suffering should not be used in war. If state actors were to use LAWS, which are lethal, then they could be in violation of this principle of necessity. As Nathan Gabriel Wood (2020, 231) argues, "until AWS are sophisticated enough to be able to competently and reliably determine what force is necessary and minimal, programing them to be able to make lethal decisions will present grave risks that *they in fact impose unnecessary or superfluous harms*" (emphasis added).

The deployment of LAWS in war could also corrupt our (less than perfect) moral nature by enabling prolonged wars we would not otherwise wage. As we have seen above, LAWS could make wars easier to wage. But LAWS cannot make dealing with the trauma of war easier. In the novel, *All Fall Down*, by James Brabazon (2020, 193), the protagonist, Max McLean, says:

> Autonomous weapons systems could tell you when to open fire *but not how to live with yourself once you had*. For now, at least, someone always had to pull the trigger or press the button—and *computers don't have nightmares about playing God* (emphasis added).

In other words, LAWS do not suffer from anxiety, stress, nightmares, or intrusive and persistent thoughts caused by trauma, including war trauma, but humans do. So, LAWS could make wars easier to wage, but we could be corrupted as a result, insofar as we would be saddled with the moral and psychological trauma of waging wars that would be easy to start but exceedingly hard to end. Consider, for instance, the ongoing tit-for-tat involving autonomous drones between Israel and Iran, including Iran's proxies, such as Hezbollah and the Houthi militia in Yemen (Amichay 2024). This (mis)use of LAWS, in turn, could lead to the devaluing of human life. As Ozlem Ulgen (2016, 10) puts it, "Autonomous weapons would devalue humanity by treating humans as disposable inanimate objects rather than ends with intrinsic value and rational thinking capacity." In that respect, the moral and psychological toll of waging wars with LAWS could be too much for less than all-good beings like us to bear.

A related moral concern is that of a LAWS arms race. Again, the Middle East provides a useful illustration of the future of warfare with LAWS here. Iran and Israel "have invested in intelligence weapons as they become more prevalent, and therefore a new AI-driven arms race may again threaten the Middle East's stability" (Sarkin and Sotoudehfar 2024, 104). Such an arms race would pose "threats to both regional and international security" (Sarkin and Sotoudehfar 2024, 102).

Like other autonomous systems, such as autonomous vehicles (see Chapter 7), LAWS are powered by machine learning algorithms that require vast amounts of data in order to function properly. In the course of their operation, LAWS would continue to collect data through their cameras, sensors, microphones, and so on, which would effectively make them rather lucrative targets for hackers, cybercriminals, and other bad actors. Even if not hacked,

LAWS could still fall into the wrong hands, such as non-state bad actors or terrorist groups, and be used to carry out terrorist attacks. According to Sarkin and Sotoudehfar (2024, 109):

> [LAWS] might become a tool more often used by non-state actors. *Their use of autonomous weapons might increase their ability to play destructive and violent roles.* This is because non-state actors, particularly terrorist groups, would be able to employ cheap autonomous drones to harm their opponents and cause chaos. The use of AI-based weapons by non-state actors will allow them to compete with the extensive military equipment of the armies of countries that are not yet equipped with this technology (emphasis added).

So, what FBI Director Christopher Wray said about autonomous vehicles (see Chapter 7), that is, that they are "a potentially new 'attack' surface for terrorists to use to harm civilians" (Straughan 2023), can be said about LAWS as well.

Finally, some LAWS, like some types of lethal autonomous drones, are relatively inexpensive such that some civilians of means may be able to purchase them. Imagine, if you will, billionaires with lethal autonomous drones, like the IAI's HARPY, or even lethal autonomous drone swarms, like those depicted in the 2017 short film *Slaughterbots*,[6] at their disposal. They might have swarms of lethal autonomous drones guarding their mansions, seeking and targeting trespassers without human intervention. This scenario is not as farfetched as it sounds. According to the Latin America and the Caribbean Human Security Network (2024):

> Using autonomous weapons as personal and private property protection, or even as a means of combating rural and urban crime, *could lead to massacres and indiscriminate killings,* as these weapons would use force against any person or object deemed a potential enemy (emphasis added).[7]

Likewise, criminal gangs could use LAWS against other gangs. "The ease and cost-effectiveness of using autonomous weapons would cause [gang] violence to spread to areas and countries where it is still precarious or non-existent, enabling criminal networks to expand their power and influence" (Latin America and the Caribbean Human Security Network 2024). Governments, too, could use LAWS as tools of repression. "Using autonomous weapons for state repression would facilitate systematic violation of human, civil, and social rights of anyone opposing the government, making it harder for victims to seek justice due to the technology's lack of traceable responsibility and accountability" (Latin America and the Caribbean Human Security Network 2024). Given their ease of use, it could be too tempting for political and military leaders to not use LAWS against their (real or perceived) enemies, political rivals, and so on. Former US President Barack Obama once joked about sending predator drones to eliminate the musical trio known as the Jonas Brothers if they "get any ideas" about his daughters (Pareene 2010).

> The Jonas Brothers are here; they're out there somewhere. Sasha and Malia are huge fans. But boys, don't get any ideas. I have two words for you, *predator drones*. You will never see it coming (emphasis added).

While President Obama was merely joking, we can easily imagine other powerful leaders who would have no moral qualms about using the lethal power of LAWS to target real or perceived enemies, political rivals, dissidents, protesters, and the like. After all, as John Emerich Edward Dalberg-Acton (1949, 364) famously said, "Power tends to corrupt and absolute power corrupts absolutely." More recently, Bendahan et al. (2015, 118) have conducted an empirical study whose findings suggest that "power affects individuals in a way that makes them behave antisocially."[8]

From all of the above, it is reasonable to conclude that LAWS could be misused or abused by (less than omnibenevolent) human beings. State actors could exploit the responsibility gaps created by LAWS to violate moral and legal conventions of war. LAWS pose significant security risks because they could be hacked by bad actors and used to carry out terrorist attacks. They could be used by criminals to engage in gang violence. Governments could use LAWS as tools for state repression, rights violations, surveillance, and assassination. Accordingly, the answer to CQ3, as far as LAWS are concerned, seems to be "yes" overall.

As far as LAWS are concerned, Critical Questions, CQ1–CQ3, seem to have affirmative answers (see Table 8.1 for a summary). This means that developing and deploying LAWS would indeed amount to "playing God" in terms of power (CQ1), knowledge (CQ2), and benevolence (CQ3). Therefore, a "playing God"

Table 8.1 Summary of the Answers (with Justifications) to the Critical Questions, CQ1–CQ3, of the "Playing God" Argument against LAWS

Critical Question	Answer	Justification
CQ1: Could LAWS get out of the control of (less than omnipotent) human beings?	yes	the control problem (AI); loss of power and control through the outsourcing of decision-making to LAWS; loss of control over responsibility when mistakes involving LAWS happen (responsibility or accountability gaps); loss of control over LAWS turning against their creators or operators
CQ2: Could LAWS be misunderstood or have unforeseen consequences for (less than omniscient) human beings?	yes	the black box problem (AI); unexplained decision-making by LAWS; unforeseen failure modes of LAWS (e.g., misidentifying targets); uncertain environmental impacts from increasing energy and water consumption to train AI for LAWS; uncertain geopolitical impacts of LAWS (e.g., increasing pace of war, instability, endless wars)
CQ3: Could LAWS be misused or abused by (less than omnibenevolent) human beings?	yes	LAWS could have a Ring of Gyges effect; LAWS could be used to violate international humanitarian law (principle of necessity); LAWS could make wars easier to wage but at great cost to our moral and psychological well-being; LAWS could be used to undermine stability and security; LAWS could be hacked; LAWS could fall into the wrong hands (e.g., terrorism); criminals could use LAWS to wage gang wars; regimes could use LAWS to repress citizens; regimes could use LAWS to facilitate rights violations; political leaders could use LAWS to assassinate political rivals

Source: Author

argument against LAWS is a strong argument (see Table 3.1 in Chapter 3). This argument can be summed up as follows.

Major Premise: We should not (morally speaking) "play God."

Conditional Premise: If LAWS could get out of the control of humans, be misunderstood or have unforeseen consequences for humans, or be misused or abused by humans, then developing and deploying LAWS amount to "playing God."

Control Premise: LAWS could get out of the control of humans.

Understanding Premise: LAWS could be misunderstood or have unforeseen consequences for humans.

Use Premise: LAWS could be misused or abused by humans.

Conclusion: We should not (morally speaking) develop and deploy LAWS.

This strong "playing God" argument against LAWS points to several moral problems with these technologies, namely, that they could get out of control, that the way they work could be misunderstood, that they could have unintended long-term consequences for us and the environment, and that they could be misused or abused by both state and non-state actors.

Like all "playing God" arguments, this one against LAWS is a defeasible argument. As we have seen in Chapter 3, conclusions of defeasible arguments can always be defeated by additional evidence that is not cited in their premises. In other words, this argument is not the final word about whether or how we should develop and deploy LAWS. One obvious way to defeat this "playing God" argument against LAWS is to argue that the potential benefits of these technologies would outweigh the perils of loss of control, misunderstanding, unforeseen consequences, misuse, and abuse. Still, this strong "playing God" argument against LAWS urges us to think carefully about the ways in which the development and deployment of LAWS could go seriously morally wrong. So, until we can make sure that we are able to control these technologies, understand them and their effects, and use them benevolently, we have a strong case against developing and deploying LAWS.

As we have seen in Chapter 3, our answers to the Critical Questions, CQ1–CQ3, will tell us what issues need to be addressed before the new or emerging technology in question can be developed, deployed, or used responsibly, that is, in accordance with our human limitations and without pretending that we are all-powerful, all-knowing, and all-good. In general, an affirmative answer to CQ1 suggests that the new or emerging technology in question could get out of the control of (less than omnipotent) human beings. Of course, we could never have perfect control over a new or emerging technology because we are not all-powerful as God is supposed to be. However, we can try to enhance our control over a new or emerging technology by analyzing risks and designing for safety. Accordingly, a "yes" to CQ1 suggests that we should direct our efforts toward risk analysis and safety. That is, we should conduct extensive risk analyses and safety testing, as well as design safety mechanisms informed by the results of these risk assessments and safety tests, to make sure that the new or emerging technology in question would be controllable by us before it can be deployed or used responsibly, that is, in accordance with our human limitations and without pretending that we are all-powerful.

An affirmative answer to CQ2 suggests that the new or emerging technology in question could be misunderstood or have unforeseen consequences for (less than omniscient) human beings. Of course, we could never have perfect understanding of a new or emerging technology and know all of its potential impacts because we are not all-knowing as God is supposed to be. However, we can try to improve our knowledge and understanding of a new or emerging technology through further research. Accordingly, a "yes" to CQ2 suggests that we should direct our efforts toward research. That is, we should conduct extensive experiments and simulations to learn as much as we can about the new or emerging technology in question before it can be deployed or used responsibly, that is, in accordance with our human limitations and without pretending that we are all-knowing.

An affirmative answer to CQ3 suggests that the new or emerging technology in question could be misused or abused by (less than omnibenevolent) human beings. Of course, we could never deploy or use a new or emerging technology in perfectly good ways because we are not all-good as God is supposed to be. However, we can try to prevent misuse and abuse by enacting regulations, policies, and laws with robust enforcement mechanisms. Accordingly, a "yes" to CQ3 suggests that we should direct our efforts toward policy. That is, we should conduct extensive deliberations, which would include all stakeholders, aimed at drafting regulations or legislating laws pertaining to the new or emerging technology in question before it can be deployed or used responsibly, that is, in accordance with our human limitations and without pretending that we are all-good.

Now, as far as the "playing God" argument against LAWS is concerned, we have answered Critical Questions CQ1–CQ3 in the affirmative. This gives us good reasons to be concerned about LAWS getting out of control, being misunderstood or having unintended, negative consequences, and being misused or abused. Having good reasons to believe this would, in turn, help us to direct our governance efforts as far as LAWS are concerned. First, our affirmative answer to CQ1 suggests that we should direct our governance efforts toward conducting risk assessments and safety tests, as well as designing safety mechanisms, including cybersecurity measures, informed by the results of these risk assessments and safety tests, for LAWS. Second, our affirmative answer to CQ2 suggests that we should direct our governance efforts toward further research on the perils of LAWS for nations, peoples, and their environments. Third, our affirmative answer to CQ3 suggests that we should direct our governance efforts toward drafting rules of war for LAWS, so as to prevent misuse and abuse by both state and non-state actors. In that respect, the "playing God" argument against LAWS, along with our answers to the Critical Questions, CQ1–CQ3, may provide some general guidance to LAWS governance.[9]

To sum up, here is how our answers to Critical Questions, CQ1–CQ3, can provide some general guidance for approaching the governance of LAWS.

CQ1: Could LAWS get out of the control of (less than omnipotent) human beings?

- Yes: attempt to enhance control by focusing on safety through risk assessment, safety testing, etc.

CQ2: Could LAWS be misunderstood or have unforeseen consequences for (less than omniscient) human beings?

- Yes: attempt to increase knowledge and understanding of consequences by focusing on research through experimentation, simulation, etc.

CQ3: Could LAWS be misused or abused by (less than omnibenevolent) human beings?

- Yes: attempt to prevent misuse or abuse by focusing on policy through regulation, legislation, enforcement, etc.

Notes

1 The video can be watched here: https://youtu.be/9CO6M2HsoIA?si=2urXq1u-1YyIzMzr.

2 Cf. Taddeo and Blanchard (2022, 1) who argue that "Only humans are morally responsible for the actions of autonomous weapons systems (AWS) [. . .] because intentions, plans, rights, and duties, praise or punishment can only be attributed in a meaningful way to humans."

3 See Wood (2023) for a taxonomy of responsibility gaps as far as autonomous weapons are concerned.

4 The video can be watched here: https://youtu.be/9CO6M2HsoIA?si=2urXq1u-1YyIzMzr.

5 Although controversial in many ways, Stanley Milgram's study on obedience to authority could be interpreted as suggesting that people are willing to act in morally reprehensible ways (e.g., delivering electric shocks of increasing intensity to another person) when they are told that someone else (e.g., an authority figure in a white lab coat) would be responsible for any wrongdoing (e.g., harm caused to a person who is receiving the electric shocks). In an updated version of Milgram's study, researchers have found evidence suggesting that "When following commands, [. . .] people genuinely feel less responsibility for their actions—whether they are told to do something evil or benign" (Abbott 2016, 394).

6 The video can be watched here: https://youtu.be/9CO6M2HsoIA?si=2urXq1u-1YyIzMzr.

7 In fact, robot dogs have been deployed to patrol Donald Trump's Mar-a-Lago resort (Jamali 2024). The news report states that these robotic dogs, which are made by Boston Dynamics, "do not have weapons" (Jamali 2024). But they can probably be weaponized without much difficulty.

8 According to Paolo Benanti, AI advisor to the United Nations and the Vatican, "There is a paternalism in Silicon Valley, a tendency to play God, meaning a desire to do good for people whether they like it or not" (Kington 2024).

9 Recall that the answers to the Critical Questions, CQ1–CQ3, of a "playing God" argument with respect to a new or emerging technology are meant to provide general guidance for approaching the governance of that technology. They are not supposed to constitute concrete governance recommendations or serve as a comprehensive governance proposal. On concrete governance proposals for the governance of LAWS, see Umbrello, Torress, and De Bellis (2020) and Chavannes, Klonowska, and Sweijs (2020).

Epilogue

Reading the news about Artificial Intelligence (AI) these days, it is difficult to avoid the sharp contrast between the hype, on the one hand, and the panic or doom, on the other hand. For example, there are those who tout the potential of AI to "revolutionize diagnosis in medicine" (Singal and Jena 2024), "boost productivity" (Brooks 2024), "fix climate," and "bring unimaginable prosperity" (Greene 2024), whereas others say that AI poses an "extinction-level threat to humans" (Egan 2024). In other words, it appears that we are given a choice between techno-optimism and techno-pessimism about AI. On the one hand, AI techno-optimist and CEO of OpenAI, Sam Altman says that "AI will solve all of our problems" (Levy 2024). On the other hand, AI techno-pessimist and cofounder of the Machine Intelligence Research Institute, Eliezer Yudkowsky says that "AI will kill us all" (Solman 2024). To avoid falling into the opposite traps of techno-optimism on the one hand and techno-pessimism on the other hand, or the hype versus doom dichotomy, we should adopt a critical attitude with respect to AI technologies. As we have seen in Chapter 1, this critical attitude toward new or emerging technologies is an intermediate position between techno-pessimism, or hype, and techno-optimism, or doom. To adopt a critical attitude with respect to new or emerging technologies, such as AI technologies, we need logical arguments.

As we have seen in Chapter 2, "playing God" arguments are frequently invoked when the morality of developing, deploying, or using a new or emerging technology is in question (Waytz and Young 2019). On my non-religious interpretation, that is, with no ontological commitment to the existence of God, first articulated in Mizrahi (2020) and further developed in Chapter 2, "playing God" arguments are not warnings to leave God's creation or nature alone, but rather invitations to think carefully about all the ways in which the development, deployment, or use of new or emerging technologies could go seriously morally wrong, given the fact that we are not all-powerful, all-knowing, and all-good as

God is supposed to be. On my interpretation, then, "playing God" arguments are critiques of humanity rather than appeals to anything supernatural like God. They are pleas to acknowledge human limitations, to avoid hubris and eventual humiliation, and to be cautious and humble instead. For this reason, on my interpretation of "playing God" arguments, they cannot be dismissed simply by raising doubts about the existence of God.

As we have seen in Chapter 3, argumentation schemes are a useful tool for the analysis and evaluation of common types of arguments that are used in everyday discourse (Walton, Reed, and Macagno 2008), such as "playing God" arguments. For this reason, the following argumentation scheme for "playing God" arguments is meant to assist in the analysis and evaluation of "playing God" arguments commonly made in everyday discourse.

"Playing God" Argumentation Scheme

Major Premise: We should not (morally speaking) "play God."
Minor Premise: Developing, deploying, or using technology *T* amounts to "playing God."
Conclusion: We should not (morally speaking) develop, deploy, or use *T*.

Accompanying this argumentation scheme for "playing God" arguments are three Critical Questions, which correspond to the divine attributes of omnipotence, omniscience, and omnibenevolence, respectively. These Critical Questions are designed to help us determine whether developing, deploying, or using a new or emerging technology would count as "playing God" when a "playing God" argument is made.

Critical Questions

CQ1: Could *T* get out of the control of (less than omnipotent) human beings?
CQ2: Could *T* be misunderstood by or have unforeseen consequences for (less than omniscient) human beings?
CQ3: Could *T* be misused or abused by (less than omnibenevolent) human beings?

These Critical Questions are conceptually related to the aforementioned scheme for "playing God" arguments in virtue of the concept of God invoked in such arguments. Unlike God, a supposedly omnipotent (all-powerful), omniscient (all-knowing), and omnibenevolent (all-good) being by definition, human beings could be overpowered, outsmarted, or overwhelmed by the power, sophistication, and potential for misuse or abuse inherent in new or emerging technologies. Moreover, CQ1–CQ3 map onto the fundamental questions at the core of many debates over the morality of new or emerging technologies. These questions are as follows.

- Can we control it? (omnipotence; CQ1)
- Do we understand it? (omniscience; CQ2)
- Will we be corrupted by it? (omnibenevolence; CQ3)

As we have seen in Chapters 1 and 2, these are fundamental questions about the limits of our power to control our environment, our knowledge of our world and our place in it, and our corruptible nature or moral character.

As we have seen in Chapter 4, my argumentation scheme for "playing God" arguments and its accompanying Critical Questions, CQ1–CQ3, are meant to be conceptual tools for the analysis and evaluation of "playing God" arguments wherever they might occur. Taken together, they can provide a constructive lens through which to determine whether we need to make sure that we are able to control, understand, and benevolently use a new or emerging technology before it is morally permissible for us to deploy and use it. To demonstrate this, I have considered four use cases for the analysis and evaluation of "playing God" arguments using my argumentation scheme for "playing God" arguments and its accompanying Critical Questions, namely, climate engineering or geoengineering (see Chapter 5), Artificial Intelligence or AI (see Chapter 6), autonomous vehicles or self-driving cars (see Chapter 7), and Lethal Autonomous Weapon Systems or LAWS (see Chapter 8). Each chapter is meant to be a standalone use case that can be used as a case study in an applied ethics course, such as Technology Ethics, Engineering Ethics, AI Ethics, and the like.

As we have seen from these use cases, the answers we give to Critical Questions, CQ1–CQ3, when a "playing God" argument about a new or emerging technology is made can help to inform our governance efforts with respect to that new or emerging technology. An affirmative answer to CQ1 suggests that we should focus our efforts on *safety* and attempt to improve our control over a new or emerging technology by conducting risk assessments and safety tests whose results should in turn inform our designing for safety. An affirmative answer to CQ2 suggests that we should focus our efforts on *research* and attempt to improve our understanding of a new or emerging technology by conducting experiments and simulations to learn as much as we can about that new or emerging technology and the consequences of deploying or using it. An affirmative answer to CQ3 suggests that we should focus our efforts on *policy* and attempt to prevent the misuse or abuse of a new or emerging technology by drafting regulations and legislating laws pertaining to the appropriate deployment or use of that new or emerging technology (see Chapter 4).

According to my interpretation, "playing God" arguments are not meant to be knockdown arguments against the development, deployment, or use of some new or emerging technology. Instead, they are defeasible arguments that are well-suited for carefully considering the perils of loss of control, misunderstanding, unforeseen consequences, misuse, and abuse of new or emerging technologies. So, far from being useless and meaningless, as some claim, "playing God" arguments allow us to assume an intermediate critical attitude between the extremes of hype and panic, boom and doom, or techno-optimism and techno-pessimism, and to approach the governance of a new or emerging technology in ways that might help to increase our control over, improve our understanding of, and prevent our misuse or abuse of new or emerging technologies.

References

Abatayo, Anna Lou, Valentina Bosetti, Marco Casari, Riccardo Ghidoni, and Massimo Tavoni (2020). "Solar Geoengineering May Lead to Excessive Cooling and High Strategic Uncertainty." *Proceedings of the National Academy of Sciences*, 117 (24): 13393–98.

Abbott, Alison (2016). "Modern Milgram Experiment Sheds Light on Power of Authority." *Nature*, 530 (7591): 394–5.

Abbott, L. F. (1924). "Radio—the New Social Force." *The Outlook*, 136 (12): 465–7.

Abney, Keith (2022). "The Ethics of Abuse and Unintended Consequences for Autonomous Vehicles." In Ryan Jenkins, David Cerný, and Tomás Hríbek (eds.), *Autonomous Vehicle Ethics: The Trolley Problem and Beyond*, 257–77. New York: Oxford University Press.

Acres, Tom (2023). "Lawyers Used ChatGPT to Help with a Case—It Backfired Massively." *Sky News*, June 23, 2023. https://news.sky.com/story/lawyers-fined-after-citing-bogus-cases-from-chatgpt-research-12908318.

Adam, David (2024). "Lethal AI Weapons are Here: How Can We Control Them?" *Nature*, 629: 521–3.

Aguirre, Anthony (2024). "The Pause Letter: One Year Later." *Future of Life Institute*, March 22, 2024. https://futureoflife.org/ai/the-pause-letter-one-year-later/.

Allyn, Bobby (2024). "More than a Dozen States Sue TikTok, Alleging It Harms Kids and Is Designed to Addict Them." *NPR*, October 8, 2024. https://www.npr.org/2024/10/08/g-s1-26823/states-sue-tiktok-child-safety-mental-health.

Allyn, Bobby, Sylvia Goodman, and Dara Kerr (2024). "States Probed TikTok for Years. Here Are the Documents the App Tried to Keep Secret." *NPR*, October 11, 2024. https://www.npr.org/2024/10/11/g-s1-27676/tiktok-redacted-documents-in-teen-safety-lawsuit-revealed.

Althorpe, Caleb, and Elizabeth Finneron-Burns (2023). "Are Savior Siblings a Special Case in Procreative Ethics?" *Journal of Ethics and Social Philosophy*, 26 (1): 49–73.

Amichay, Rami (2024). "Tel Aviv Hit by Drone Attack Claimed by Iranian-Backed Houthis." *Reuters*, July 19, 2024. https://www.reuters.com/world/middle-east/israeli-military-says-tel-aviv-blast-apparently-caused-by-drone-2024-07-19/.

Amoroso, Daniele, and Guglielmo Tamburrini (2018). "The Ethical and Legal Case against Autonomy in Weapons Systems." *Global Jurist*, 18 (1): 1934–2640.

Amoroso, Daniele, and Guglielmo Tamburrini (2020). "Autonomous Weapons Systems and Meaningful Human Control: Ethical and Legal Issues." *Current Robotics Reports*, 1: 187–94.

Andreessen, Marc (2023). "The Techno-Optimist Manifesto." *Andreessen Horowitz*, October 16, 2023. https://a16z.com/the-techno-optimist-manifesto/.

Anselm (1995). *Proslogion with the Replies of Gaunilo and Anselm*, trans. Thomas Williams. Indianapolis: Hackett Publishing Company, Inc.

Appleton, Evan, Kyunghee Hong, Cristina Rodríguez-Caycedo, Yoshiaki Tanaka, Asaf Ashkenazy-Titelman, Ketaki Bhide, Cody Rasmussen-Ivey, Xochitl Ambriz-Peña, Nataly Korover, and Hao Bai (2024). "Derivation of Elephant Induced Pluripotent Stem Cells." *bioRxiv*, March 12, 2024. https://doi.org/10.1101/2024.03.05.583606.

Aristotle (1984). *The Complete Works of Aristotle: The Revised Oxford Translation*, vols 1 and 2, ed. Jonathan Barnes. Princeton, NJ: Princeton University Press.

Atomic Heritage Foundation (2022). "Marhsall Islands." *National Museum of Nuclear Science and History*. https://ahf.nuclearmuseum.org/ahf/location/marshall-islands/ (accessed September 2, 2024).

Baer, Hans A. (2019). *Motor Vehicles, the Environment, and the Human Condition: Driving to Extinction*. London: Rowman & Littlefield.

Ball, Philip (2010). "'Playing God' is a Meaningless, Dangerous Cliché." *Prospect*, May 24, 2010. https://www.prospectmagazine.co.uk/science-and-technology/playing-god-is-a-meaningless-dangerous-cliche.

Barry, Vincent (2012). *Bioethics in a Cultural Context: Philosophy, Religion, History, Politics*. Boston, MA: Wadsworth.

Battista, Daniele (2024). "Political Communication in the Age of Artificial Intelligence: An Overview of Deepfakes and Their Implications." *Society Register*, 8 (2): 7–24.

Baum, Seth, D., Timothy M. Maher Jr., and Jacob Haqq-Misra (2013). "Double Catastrophe: Intermittent Stratospheric Geoengineering Induced by Societal Collapse." *Environment Systems & Decisions*, 33 (1): 168–80.

Bay, Bjørn, Erik Lykke Mortensen, and Ulrik Schiøler Kesmodel (2013). "Assisted Reproduction and Child Neurodevelopmental Outcomes: A Systematic Review." *Fertility and Sterility*, 100 (3): 844–53.

Bendahan, Samuel, Christian Zehnder, François P. Pralong, and John Antonakis (2015). "Leader Corruption Depends on Power and Testosterone." *The Leadership Quarterly*, 26 (2): 101–22.

Berry, Roberta M. (2007). *The Ethics of Genetic Engineering*. London: Routledge.

Bhalla, Nita, and Mansi Thapliyal (2013). "India Seeks to Regulate Its Booming 'Rent-a-Womb' Industry." *Reuters*, September 30, 2013. https://www.reuters.com/article/us-india-surrogates-idUSBRE98T07F20130930/.

Biermann, Frank, and Ina Möller (2019). "Rich Man's Solution? Climate Engineering Discourses and the Marginalization of the Global South." *International Environmental Agreements: Politics, Law, and Economics*, 19 (2): 151–67.

Boafo, James, Jacob Obodai, Eric Stemn, and Philip Nti Nkrumah (2024). "The Race for Critical Minerals in Africa: A Blessing or Another Resource Curse?" *Resources Policy*, 93: 105046.

Boddington, Paula (2017). *Towards a Code of Ethics for Artificial Intelligence*. Gewerbestrasse: Springer.

Boddington, Paula (2023). *AI Ethics: A Textbook*. Singapore: Springer.

Bond, Shannon (2024). "AI Fakes Raise Election Risks as Lawmakers and Tech Companies Scramble to Catch Up." *NPR*, February 8, 2024. https://www.npr.org/2024/02/08/1229641751/ai-deepfakes-election-risks-lawmakers-tech-companies-artificial-intelligence.

Brabazon, James (2020). *All Fall Down*. Berkely: Penguin Random House.

Brooks, Chad (2024). "AI Boosts Small Business Productivity, But Employee Training Lags Behind." *business.com*, September 16, 2024. https://www.business.com/articles/ai-usage-smb-workplace-study/.

Buckwalter, Wesley (2019). "Theoretical Motivation of 'Ought Implies Can.'" *Philosophia*, 48 (1): 83–94.

Burgess, Matt (2023). "Deepfake Porn is Out of Control." *Wired*, October 16, 2023. https://www.wired.com/story/deepfake-porn-is-out-of-control/.

Campaign to Stop Killer Robots (2013). "Urgent Action Needed to Ban Fully Autonomous Weapons." *Campaign to Stop Killer Robots*, April 23, 2013. https://www.stopkillerrobots.org/wp-content/uploads/2013/04/KRC_LaunchStatement_23Apr2013.pdf.

Cavanaugh, Thomas A. (2002). "'Playing God' and Bioethics." *Christian Bioethics: Non-Ecumenical Studies in Medical Morality*, 8 (2): 119–24.

Cavero-Ibiricu, Amaia, Javier Canelas-Fernández, Inés Gómez-Acebo, Jessica Alonso-Molero, Daniel Martínez-Jiménez, Javier Llorca, María J. Cabero-Perez, and Trinidad Dierssen-Sotos (2024). "Association between Assisted Reproductive Technology and Cerebral Palsy: A Meta-Analysis." *Pediatric Neurology*, 152: 115–24.

Cech, Laura (2022). "AI to Detect Sepsis." *Johns Hopkins Magazine*, Winter 2022. https://hub.jhu.edu/magazine/2022/winter/ai-technology-to-detect-sepsis/.

Centers for Disease Control and Prevention (2023). "2021 Assisted Reproductive Technology Fertility Clinic and National Summery Report." *US Department of Health and Human Services*. https://www.cdc.gov/art/reports/2021/pdf/Report-ART-Fertility-Clinic-National-Summary-H.pdf (accessed August 21, 2024).

Chavannes, Esther, Klaudia Klonowska, and Tim Sweijs (2020). *Governing Autonomous Weapon Systems: Expanding the Solution Space, From Scoping to Applying*. The Hague Centre for Strategic Studies. https://hcss.nl/wp-content/uploads/2021/01/HCSS-Governing-AWS-final.pdf.

Cheng, Cecilia, Yan-Ching Lau, Linus Chan, and Jeremy W. Luk (2021). "Prevalence of Social Media Addiction Across 32 Nations: Meta-analysis with Subgroup Analysis of Classification Schemes and Cultural Values." *Addictive Behaviors*, 117: 106845.

Chillingsworth, Luke (2023). "I'm a Car Expert – Driverless Cars Come with These Concerning Risks." *Daily Express*, October 28, 2023. https://www.express.co.uk/life-style/cars/1828999/driverless-cars-risks-safety-autonomous-vehicles.

Choi, Charles Q. (2019). "How Self-Driving Cars Might Transform City Parking." *IEEE Spectrum*, February 20, 2019. https://spectrum.ieee.org/autonomous-parking.

Cohen, Jon (2022). "Where Does Embryo Editing Stand?" *Science*, 375 (6587): 1328–9.

Cohen, Rachel, Toby Newton-John, and Amy Slater (2018). "'Selfie'-objectification: The Role of Selfies in Self-objectification and Disordered Eating in Young Women." *Computers in Human Behavior*, 79: 68–74.

Corner, Adam, and Nick Pidgeon (2014). "Geoengineering, Climate Change Scepticism and the 'Moral Hazard' Argument: An Experimental Study of UK Public Perceptions." *Philosophical Transactions of the Royal Society*, 372 (2031): 20140063.

Corvino, John (2013). *What's Wrong with Homosexuality?* New York: Oxford University Press.

C-SPAN (2023). "World Economic Forum Discussion on Technology and National Security." *C-SPAN*, January 19, 2023. https://www.c-span.org/video/?525429-1/world-economic-forum-discussion-technology-national-security.

Cummings, Mary L. (2023). "What Self-Driving Cars Tell Us About AI Risks." *IEEE Spectrum*, July 30, 2023. https://spectrum.ieee.org/self-driving-cars-2662494269.

Dalberg-Acton, Edward Emerich John (1949). *Essays on Freedom and Power*. Glencoe, IL: The Free Press.

Danaher, John (2016). "The Threat of Algocracy: Reality, Resistance and Accommodation." *Philosophy & Technology*, 29: 245–68.

Danaher, John, and Jeroen Hopster (2022). "The Normative Significance of Future Moral Revolutions." *Futures*, 144: 103046.

De Bie, Felix R., Sarah D. Kim, Sourav K. Bose, Pamela Nathanson, Emily A. Partridge, Alan W. Flake, and Chris Feudtner (2023). "Ethics Considerations

Regarding Artificial Womb Technology for the Fetonate." *The American Journal of Bioethics*, 23 (5): 67–78.

Diplomatic Conference on the Reaffirmation and Development of International Humanitarian Law Applicable in Armed Conflicts (1977). "Protocol Additional to the Geneva Conventions of 12 August 1949, and Relating to the Protection of Victims of International Armed Conflicts." *United Nations Office of the High Commissioner*, June 8, 1977. https://www.ohchr.org/en/instruments-mechanisms/instruments/protocol-additional-geneva-conventions-12-august-1949-and.

Dresp-Langley, Birgitta (2023). "The Weaponization of Artificial Intelligence: What the Public Needs to be Aware of." *Frontiers in Artificial Intelligence*, 6: 1154184.

Duarte, Tania, Nicholas Barrow, Medina Bakayeva, and Peter Smith (2024). "Editorial: The Ethical Implications of AI Hype." *AI and Ethics*, 4 (3): 649–51.

Eady, Gregory, Tom Paskhalis, Jan Zilinsky, Richard Bonneau, Jonathan Nagler, and Joshua A. Tucker (2023). "Exposure to the Russian Internet Research Agency Foreign Influence Campaign on Twitter in the 2016 US Election and Its Relationship to Attitudes and Voting Behavior." *Nature Communications*, 14: 62.

Egan, Matt (2024). "AI Could Pose 'Extinction-Level' Threat to Humans and the US Must Intervene, State Dept.-Commissioned Reports Warns." *CNN*, March 12, 2024. https://www.cnn.com/2024/03/12/business/artificial-intelligence-ai-report-extinction/index.html.

Etzioni, Oren (2018). "A Hippocratic Oath for Artificial Intelligence Practitioners." *TechCrunch*, March 14, 2018. https://techcrunch.com/2018/03/14/a-hippocratic-oath-for-artificial-intelligence-practitioners/.

European Parliament (2024). "EU AI Act: First Regulation on Artificial Intelligence." *European Parliament*, June 18, 2024. https://www.europarl.europa.eu/topics/en/article/20230601STO93804/eu-ai-act-first-regulation-on-artificial-intelligence.

Evans, John H. (2002). *Playing God? Human Genetic Engineering and the Rationalization of Public Bioethical Debate*. Chicago: University of Chicago Press.

Federal Reserve Banks (2019). "Payments Fraud Insights: Synthetic Identity Fraud in the U.S. Payment System." *The Federal Reserve FedPayments Improvement*, July 2019. https://fedpaymentsimprovement.org/wp-content/uploads/frs-synthetic-identity-payments-fraud-white-paper-july-2019.pdf.

Feng, Emily, and Lexie Schapitl (2023). "How a Chinese 'Spy Balloon' Prompted the U.S. to Scour the Skies." *NPR*, February 14, 2023. https://www.npr.org/2023/02/14/1156731462/china-spy-balloon-timeline-key-dates.

Fleischer, Will (2022). "Understanding, Idealization, and Explainable AI." *Episteme*, 19 (4): 534–60.

Fletcher, Emma (2023). "Social Media: A Golden Goose for Scammers." *Federal Trade Commission, Consumer Protection*, October 6, 2023. https://www.ftc.gov/news-events/data-visualizations/data-spotlight/2023/10/social-media-golden-goose-scammers#1.

Floridi, Luciano (2023). *The Ethics of Artificial Intelligence*. New York: Oxford University Press.

Funk, Cary, and Meg Hefferon (2018). "Most Americans Accept Genetic Engineering of Animals that Benefits Human Health, but Many Oppose Other Uses." *Pew Research Center*, August 16, 2018. https://www.pewresearch.org/science/2018/08/16/most-americans-accept-genetic-engineering-of-animals-that-benefits-human-health-but-many-oppose-other-uses/.

Gantsho, Luvuyo (2022). "God Does not Play Dice but Self-Driving Cars Should." *AI and Ethics*, 2 (1): 177–84.

Gardiner, Stephen, and Augustine Fragnière (2018). "The Tollgate Principles for the Governance of Geoengineering: Moving Beyond the Oxford Principles to an Ethically more Robust Approach." *Ethics, Policy, & Environment*, 21 (2): 143–74.

Garrido-Rodriguez, Pedro, Miguel Carmena-Bargueño, Maria Eugenia de la Morena-Barrio, Carlos Bravo-Perez, Belen de la Morena-Barrio, Rosa Cifuentes-Riquelme, MariaLuisa Lozano, Horacio Perez-Sanchez, and Javier Corral (2024). "Analysis of AlphaFold and Molecular Dynamics Structure Predictions of Mutations in Serpins." *PLoS ONE*, 19 (7): 0304451. https://doi.org/10.1371/journal.pone.0304451.

Gecker, Jocelyn (2023). "Paper Exams, Chatbot Bans: Colleges Seek to 'ChatGPT-Proof' Assignments." *The Associated Press*, August 10, 2023. https://apnews.com/article/chatgpt-cheating-ai-college-1b654b44de2d0dfa4e50bf0186137fc1.

Graham, Gordon (2002). *Genes: A Philosophical Inquiry*. London: Routledge.

Green, Brian P. (2022). *Space Ethics*. London: Rowman & Littlefield.

Greene, Tristan (2024). "Sam Altman Says ChatGPT Will Bring Unimaginable Prosperity, Fix Climate." *Cointelgraph*, September 23, 2024. https://cointelegraph.com/news/sam-altman-chatgpt-unimaginable-prosperity-fix-climate.

Gridneff, Ilya, Emily Schultheis, and Dmytro Drabyk (2023). "Inside a Ukrainian Baby Factory." *Politico*, July 23, 2023. https://www.politico.com/news/2023/07/23/ukraine-surrogates-fertility-00104913.

Gubrud, Mark (2016). "Why Should We Ban Autonomous Weapons? To Survive." *IEEE Spectrum*, June 1, 2016. https://spectrum.ieee.org/why-should-we-ban-autonomous-weapons-to-survive.

Hall, Shannon (2015). "Recklessly Playing God with the Planet." *Science Line*, June 3, 2015. https://scienceline.org/2015/06/recklessly-playing-god-with-the-planet/.

Hamblin, James (2015). "A Brewing Problem: Wha's the Healthiest Way to Keep Everyone Caffeinated?" *The Atlantic*, March 2, 2015. https://www.theatlantic.com/technology/archive/2015/03/the-abominable-k-cup-coffee-pod-environment-problem/386501/.

Hambling, David (2021). "Israel Used World's First AI-Guided Combat Drone Swarm in Gaza Attacks." *New Scientist*, June 30, 2021. https://www.newscientist.com/article/2282656-israel-used-worlds-first-ai-guided-combat-drone-swarm-in-gaza-attacks/.

Hambling, David (2023). "US Air Force Is Giving Military Drones the Ability to Recognise Faces." *New Scientist*, February 23, 2023. https://www.newscientist.com/article/2360475-us-air-force-is-giving-military-drones-the-ability-to-recognise-faces/.

Hamilton, Clive (2013). *Earthmasters: The Dawn of the Age of Climate Engineering.* New Have: Yale University Press.

Hand to Hold (2017). "Could This Artificial Womb Revolutionize Micro Preemie Care?" *Hand to Hold Blog*, May 4, 2017. https://handtohold.org/artificial-womb/.

Hanif, Muhammad Shahid (2024). "Synthetic Identities: The Darker Side of Generative AI." *Forbes*, May 29, 2024. https://www.forbes.com/councils/forbestechcouncil/2024/05/29/synthetic-identities-the-darker-side-of-generative-ai/.

Hansson, Sven Ove, Matts-Åke Belin, and Björn Lundgren. 2021. "Self-Driving Vehicles—An Ethical Overview." *Philosophy & Technology*, 34: 1383–408.

Hao, Karen (2019). "A US Government Study Confirms Most Face Recognition Systems are Racist." *MIT Technology Review*, December 20, 2019. https://www.technologyreview.com/2019/12/20/79/ai-face-recognition-racist-us-government-nist-study/.

Hartman, Laura M. (2017). "Climate Engineering and the Playing God Critique." *Ethics & International Affairs*, 31 (3): 313–33.

Hassija, Vikas, Vinay Chamola, Atmesh Mahapatra, Abhinandan Singal, Divyansh Goel, Kaizhu Huang, Simone Scadapane, Indro Spinelli, Mufti Mahmud, and Amir Hussain (2024). "Interpreting Black-Box Models: A Review on Explainable Artificial Intelligence." *Cognitive Computation*, 16: 45–74. https://doi.org/10.1007/s12559-023-10179-8.

Hawkins, Andrew J. (2018). "For the Love of God, Don't Play Chicken with the Self-Driving Cars: Smart Cars vs. Dumb Meatsacks." *The Verge*, January 16, 2018. https://www.theverge.com/2018/1/16/16896272/self-driving-car-human-interaction-waymo.

Hawkins, Andrew J. (2024). "Tesla Cybercab Announced: Elon Musk's Robotaxi is Finally Here." *The Verge*, October 10, 2024. https://www.theverge.com/2024/10/10/24265530/tesla-robotaxi-elon-musk-features-range-price-release-date.

Herzfeld, Noreen (2017). "Introduction: Religion and the New Technologies." In Noreen Herzfeld (ed.), *Religion and the New Technologies*, 1–3. Basel: MDPI.

Hoel, Erik (2024). "A.I.-Generated Garbage is Polluting Our Culture." *The New York Times*, March 29, 2024. https://www.nytimes.com/2024/03/29/opinion/ai-internet-x-youtube.html.

Hoffman, Fabian, and Graham Feingold (2021). "Cloud Microphysical Implications for Marine Cloud Brightening: The Importance of the Seeded Particle Size Distribution." *Journal of the Atmospheric Sciences*, 78 (1): 3247–62.

Hoffman, Joshua, and Gary Rosenkrantz (2013). "Omnipotence." In Chad Meister and Paul Copan (eds.), *The Routledge Companion to Philosophy of Religion*, 319–28. New York: Routledge.

Hoffman, Joshua, and Gary Rosenkrantz (2017). "Omnipotence." In Edward N. Zalta (ed.), *Stanford Encyclopedia of Philosophy*, Winter 2017 Edition. https://plato.stanford.edu/archives/win2017/entries/omnipotence//

Hsu, Jeremy (2024). "Driverless Cars are Mostly Safer than Humans—But Worse at Turns." *New Scientist*, June 18, 2024. https://www.newscientist.com/article/2435896-driverless-cars-are-mostly-safer-than-humans-but-worse-at-turns/.

Hughes, Trevor, and Minnah Arshad (2024). "Alabama Justice Invoked 'the Wrath of a Holy God' in IVF Opinion. Is that Allowed?" *USA Today*, February 22, 2024. https://www.usatoday.com/story/news/nation/2024/02/22/ivf-opinion-from-alabama-justice-was-overtly-religious/72689378007/.

Human Rights Watch (2012). "Losing Humanity: The Case Against Killer Robots." *Human Rights Watch*, November 19, 2012. https://www.hrw.org/report/2012/11/19/losing-humanity/case-against-killer-robots.

Humphreys, Declan, Abigail Koay, Dennis Desmond, and Erica Mealy (2024). "AI Hype as a Cyber Security Risk: The Moral Responsibility of Implementing Generative AI in Business." *AI and Ethics*, 4 (3): 791–804.

Hysolli, Eriona, Elizabeth A. Hadly, and Deborah A. Sivas (2023). "Should We Bring Species Back from Extinction?" *The Nation*, May 26, 2023. https://www.thenation.com/article/environment/de-extinction-debate/.

IAI (2023). "HARPY: Anti-Radiation Loitering Munition." *IAI*. https://www.iai.co.il/p/harpy (accessed September 16, 2024).

IEA (2024). "Electricity 2024: Analysis and Forecast to 2026." Paris: IEA. https://www.iea.org/reports/electricity-2024.

Ingram, David (2023). "Driver Hits Pedestrian, Pushing Her into Path of Self-Driving Car in San Fransisco." *NBC News*, October 3, 2023. https://www.nbcnews.com/tech/tech-news/driver-hits-pedestrian-pushing-path-self-driving-car-san-francisco-rcna118603.

Inhorn, Marcia C., and Pasquale Patrizio (2009). "Rethinking Reproductive 'Tourism' as Reproductive 'Exile.'" *Fertility and Sterility*, 92 (3): 904–6.

IPCC (2023). "Climate Change 2023: Synthesis Report." Contribution of Working Groups I, II and III to the Sixth Assessment Report of the Intergovernmental Panel on Climate Change. IPCC, Geneva, Switzerland. https://www.ipcc.ch/report/ar6/syr/downloads/report/IPCC_AR6_SYR_FullVolume.pdf.

Isaacson, Walter (2016). "How to Fix the Internet: Anonymity Has Poisoned Online Life." *The Atlantic*, December 15, 2016. https://www.theatlantic.com/technology/archive/2016/12/how-to-fix-the-internet/510797/.

Jamali, Lily (2024). "'Do Not Pet': Why are Robot Dogs Patrolling Mar-A-Lago?" *BBC*, November 17, 2024. https://www.bbc.com/news/articles/c30p16gn3pvo.

Johnson, Courtney, and Brian Kennedy (2021). "U.S. Adults have Mixed Views on whether Geoengineering would Help Reduce Effects of Climate Change." *Pew Research Center*, June 11, 2021. https://www.pewresearch.org/short-reads/2021/06/11/u-s-adults-have-mixed-views-on-whether-geoengineering-would-help-reduce-effects-of-climate-change/.

Jones, Ernest (1923). *Essays in Applied Psycho-Analysis*. London: The International Psycho-Analytical Press.

Kallenborn, Zachary (2024). "Swarm Clouds on the Horizon? Exploring the Future of Drone Sarm Proliferation." *Modern War Institute at West Point*, March 20, 2024. https://mwi.westpoint.edu/swarm-clouds-on-the-horizon-exploring-the-future-of-drone-swarm-proliferation/.

Kamenjasevic, Erik (2024). "Ethical and Legal Considerations of Mood Enhancement Technology." *Frontiers in Human Dynamics*, 6: 1310144.

Kass, Leon R. (1997). "The Wisdom of Repugnance: Why We Should Ban the Cloning of Humans." *The New Republic*, June 2, 1997. https://web.stanford.edu/~mvr2j/sfsu09/extra/Kass2.pdf.

Kington, Tom (2024). "Stop Playing God, Pope's Adviser Tells Tech Titans." *The Times*, January 19, 2024. https://www.thetimes.com/world/article/stop-playing-god-popes-adviser-tells-tech-titans-pfvn3jp5m.

Kissinger-Knox, Alicia, Patrick Aragon, and Moti Mizrahi (2018). "'Ought Implies Can', Framing Effects, and 'Empirical Refutations.'" *Philosophia*, 46 (1): 165–82.

Klaas, Brian (2024). *Fluke: Chance, Chaos, and Why Everything We Do Matters*. New York: Scribner.

Klitzman, Robert, and Mark V. Sauer (2015). "Creating and Selling Embryos for 'Donation': Ethical Challenges." *American Journal of Obstetrics and Gynecology*, 212 (2): 167–70.

Knowles, Tom (2024). "Meet the Ordinary People Whose Lives Have been Ruined by AI." *The Telegraph*, March 13, 2024. https://www.telegraph.co.uk/books/non-fiction/review-code-dependent-living-shadow-ai-madhumita-murgia/.

Koren, Marina (2024). "Playing God with the Atmosphere." *The Atlantic*, April 18, 2024. https://www.theatlantic.com/science/archive/2024/04/dubai-oman-flooding-cloud-seeding-geoengineering/678114/.

Kraft, Dina (2011). "Where Families are Prized, Help is Free." *The New York Times*, July 17, 2011. https://www.nytimes.com/2011/07/18/world/middleeast/18israel.html.

Krakovna, Victoria, Meita Chita-Tegmark, Max Tegmark, Jaan Tallinn, and Anthony Aguirre (2023). "Pause Giant AI Experiments: An Open Letter." *Future of Life Institute*, March 22, 2023. https://futureoflife.org/open-letter/pause-giant-ai-experiments/.

La Fors, Karolina (2020). "Legal Remedies for a Forgiving Society: Children's Rights, Data Protection Rights and the Value of Forgiveness in AI-mediated Risk Profiling of Children by Dutch Authorities." *Computer Law & Security Review*, 38: 105430.

Latin America and the Caribbean Human Security Network (2024). "The Dangers of Autonomous Weapons Systems from a Latin American Perspective." *United Nations General Assembly First Committee, Replies to the Secretary-General*, May 24, 2024. https://docs-library.unoda.org/General_Assembly_First_Committee_-Seventy-Ninth_session_(2024)/78-241-SEHLAC-EN.pdf.

Lawrence, Mark G., Stefan Schäfer, Helene Muri, Vivian Scott, Andreas Oschilles, Naomi E. Vaughan, Olivier Boucher, Hauke Schmidt, Jim Haywood, and Jürgen Scheffran (2018). "Evaluating Climate Geoengineering Proposals in the Context of the Paris Agreement Temperature Goals." *Nature Communications*, 9: 3734.

Lee, Sun Kyong, Juhyung Sun, Seulki Jang, and Shane Connelly (2022). "Misinformation of COVID-19 Vaccines and Vaccine Hesitancy." *Scientific Reports*, 12: 13681.

Levy, Steven (2024). "No, Sam Altman, AI Won't Solve all of Humanity's Problems." *Wired*, September 27, 2024. https://www.wired.com/story/plaintext-no-sam-altman-ai-wont-solve-all-of-humanitys-problems/.

Li, Pengfei, Jianyi Yang, Mohammad A. Islam, and Shaolei Ren (2023). "Making AI less 'Thirsty': Uncovering and Addressing the Secret Water Footprint of AI Models." https://arxiv.org/abs/2304.03271v3.

Link, Hans-Jürgen (2013). "Playing God and the Intrinsic Value of Life: Moral Problems for Synthetic Biology?" *Science and Engineering Ethics*, 19 (2): 435–48.

Liou, Tiffany (2024). "'I don't Want to Live in Fear Anymore': North Texas Girl Victimized with Deepfake Nudes Pushes for Federal Law." *WFAA*, June 21,

2024. https://www.wfaa.com/article/news/local/north-texas-girl-victimized-with-deepfake-nudes-pushes-for-federal-law/287-7cf49849-1b27-4a21-864c-d69cf1644e9c.

Luppicini, Rocci (2012). *Ethical Impact of Technological Advancements and Applications in Society*. Hershey, PA: IGI Global.

Lynas, Mark (2011). "Geoengineering, Nuclear Power and Climate Change: Playing God is Good for the Planet." *The Telegraph*, July 12, 2011. https://www.telegraph.co.uk/news/science/science-news/8631604/Geo-engineering-nuclear-power-and-climate-change-playing-God-is-good-for-the-planet.html.

Määttänen, Anni, Thibault Lameille, Carola Kloeck, Olivier Boucher, and François Ravetta (2024). "Uncertainties and Confidence in Stratospheric Aerosol Injection Modelling: A Systematic Literature Review." *Oxford Open Climate Change*, 4 (1): kgae007.

Marshall, Aarian (2022). "Cruise's Robot Car Outages Are Jamming Up San Fransisco." *Wired*, July 8, 2022. https://www.wired.com/story/cruises-robot-car-outages/.

McSweeney, Robert (2019). "The Carbon Brief Interview: Prof Joanna Haigh." *Carbon Brief*, April 29, 2019. https://www.carbonbrief.org/the-carbon-brief-interview-prof-joanna-haigh/.

Mellor, Duane (2024). "The Role of Food Science and Technology in Navigating the Health Issues of Ultra-Processed Foods." *International Journal of Food Science and Technology*, 59 (9): 6688–92.

Menninger, Karl (1959). "Verdict Guilty—Now What?" *The Pointer News*, 111 (3): 12–15.

Middleburg, K. J., M. J. Heineman, A. F. Bos, and M. Hadders-Algra (2008). "Neuromotor, Cognitive, Language and Behavioral Outcome in Children Born Following IVF or ICSI – A Systematic Review." *Human Reproduction Update*, 14 (3): 219–31.

Mishra, Ashutosh (2015). "Impact of Silica Mining on Environment." *Journal of Geography and Regional Planning*, 8 (6): 150–6.

Mizrahi, Moti (2013). "New Puzzles about Divine Attributes." *European Journal for Philosophy of Religion*, 5 (2): 147–57.

Mizrahi, Moti (ed.) (2018). *The Kuhnian Image of Science: Time for a Decisive Transformation?* London: Rowman & Littlefield Publishers.

Mizrahi, Moti (2020). "How to Play the 'Playing God' Card." *Science and Engineering Ethics*, 26 (3): 1445–61.

Mizrahi, Moti (2020). *The Relativity of Theory: Key Positions and Arguments in the Contemporary Scientific Realism/Antirealism Debate*. Cham, Switzerland: Springer.

Mizrahi, Moti (ed.) (2022). *For and Against Scientism: Science, Methodology, and the Future of Philosophy*. Lanham: Rowman & Littlefield Publishers.

Mizrahi, Moti (2023). "Should Punishment Fit the Crime? Dr. Karl Menninger on the Crime of Punishment." *JSTOR Daily*, June 15, 2023. https://daily.jstor.org/should-punishment-fit-the-crime/.

Morrow, David R. (2020). *Values in Climate Policy*. London: Rowman & Littlefield.

Nam, Ju Gang, Sunggyun Park, Eui Jin Hwang, Jong Hyuk Lee, Kwang-Nam Jin, Kun Young Lim, Thienkai Huy Vu, Jae Ho Sohn, Sangheum Hwang, Jin Mo Goo, and Chang Min Park (2018). "Development and Validation of Deep Learning-Based Automatic Detection Algorithm for Malignant Pulmonary Nodules on Chest Radiographs." *Radiology*, 290 (1): 218–28.

Ndasauka, Yamikani, Zhengde Wei, and Xiaochu Zhang (2017). "Received View of Addiction, Relapse, and Treatment." In Xiaochu Zhang, Jie Shi, and Ran Tao (eds.), *Substance and Non-Substance Addiction*, 3–20. Gateway East: Springer.

Nicastro, Dom (2024). "Sam Altman: AI Will Replace 95% of Creative Marketing Work." *CMSWire*, March 13, 2024. https://www.cmswire.com/digital-marketing/sam-altman-ai-will-replace-95-of-creative-marketing-work/.

Nugent, Ciara (2018). "What it was Like to Grow Up as the World's First 'Test-Tube Baby.'" *Time*, July 25, 2018. https://time.com/5344145/louise-brown-test-tube-baby/.

O'Brien, Matt, and Hannah Fingerhut (2023). "Artificial Intelligence Technology Behind ChatGPT was Built in Iowa—With a Lot of Water." *The Associated Press*, September 9, 2023. https://apnews.com/article/chatgpt-gpt4-iowa-ai-water-consumption-microsoft-f551fde98083d17a7e8d904f8be822c4.

OECD (2024). "Recommendation of the Council on Artificial Intelligence." *OECD Legal Instruments* 0049. file:///C:/Users/mmizrahi/Downloads/OECD-LEGAL-0449-en.pdf.

Office of the Under Secretary of Defense for Policy (2023). "DoD Directive 3000.09, Autonomy in Weapon Systems." *Department of Defense*, January 25, 2023. https://media.defense.gov/2023/Jan/25/2003149928/-1/-1/0/DOD-DIRECTIVE-3000.09-AUTONOMY-IN-WEAPON-SYSTEMS.PDF.

Oimann, Ann-Katrien (2023). "The Responsibility Gap and LAWS: A Critical Mapping of the Debate." *Philosophy & Technology*, 36 (1): 1–22.

Oppy, Graham (2006). *Arguing about Gods*. New York: Cambridge University Press.

Orben, Amy (2020). "The Sisyphean Cycle of Technology Panics." *Perspectives on Psychological Science*, 15 (5): 1143–57.

Ortutay, Barbara (2023). "States Sue Meta Claiming its Social Platforms are Addictive and Harm Children's Mental Health." *The Associated Press*, October 24, 2023.

https://apnews.com/article/instagram-facebook-children-teens-harms-lawsuit-attorney-general-1805492a38f7cee111cbb865cc786c28.

Pareene, Alex (2010). "Obama Threatens Jonas Brothers with Drone Strikes." *Salon*, May 3, 2010. https://www.salon.com/2010/05/03/obama_drone_joke_jonas_brothers/.

Parker, Andy, and Peter J. Irvine (2018). "The Risk of Termination Shock from Solar Geoengineering." *Earth's Future*, 6: 456–67.

Pelton, Joseph N. (2013). *Space Debris and Other Threats from Outer Space*. Dordrecht: Springer.

Peters, Ted (2003). *Playing God? Genetic Determinism and Human Freedom*. 2nd ed. New York: Routledge.

Peters, Ted (2007). "Are We Playing God with Nanoenhancement?" In Fritz Allhoff, Patrick Lin, James Moor, and John Weckert (eds.), *Nanoethics: The Ethical and Social Implications of Nanotechnology*, 173–84. Hoboken, NJ: Wiley & Sons, Inc.

Peterson, Martin (2020). *Ethics for Engineers*. New York: Oxford University Press.

Piersanti, Valeria, Francesca Consalvo, Fabrizio Signore, Allesandro Del Rio, and Simona Zaami (2021). "Surrogacy and 'Procreative Tourism': What Does the Future Hold from the Ethical and Legal Perspectives? *Medicina (Kaunas)*, 57 (1): 47.

Prentice, Chris, and Susan Heavey (2024). "US SEC hits Keurig with $1.5 mln Fine Over Pod Recyclability Statements." *Reuters*, September 10, 2024. https://www.reuters.com/business/retail-consumer/keurig-pay-15mln-fine-over-inaccurate-pod-recyclability-statements-sec-says-2024-09-10/.

Proctor, Jonathan, Solomon Hsiang, Jennifer Burney, Marshall Burke, and Wolfram Schlenker (2018). "Estimating Global Agricultural Effects of Geoengineering Using Volcanic Eruptions." *Nature*, 560: 480–3.

Ramsey, Paul (1970). *Fabricated Man: The Ethics of Genetic Control*. New Haven: Yale University Press.

Ravitsky, Vardit, and Louise King (2024). "It Is Too Soon for Clinical Trials on Artificial Wombs." *Scientific American*, June 1, 2024. https://www.scientificamerican.com/article/it-is-too-soon-for-clinical-trials-on-artificial-wombs/.

Rayner, Steve, Clare Heyward, Tim Kruger, Nick Pidgeon, Catherine Redgwell, and Julian Savulescu (2013). "The Oxford Principles." *Climate Change*, 121: 499–512.

Relevant (2022). "An 'Artificial Womb Facility' that Gives Parents the Ability to Pick their Child's Traits Is in the Works." *Relevant*, December 12, 2022. https://relevantmagazine.com/culture/tech-gaming/an-artificial-womb-facility-that-gives-parents-the-ability-to-pick-their-childs-traits-is-in-the-works/.

Roberts, Huw, Emmie Hine, Mariarosaria Taddeo, and Luciano Floridi (2024). "Global AI Governance: Barriers and Pathways Forward." *International Affairs*, 100 (3): 1275–86.

Robock, Alan (2008). "20 Reasons Why Geoengineering May be a Bad Idea." *Bulletin of the Atomic Scientists*, 64 (2): 14–18.

Robock, Alan, Martin Bunzi, Ben Kravitz, and Georgiy L. Stenchikov (2010). "A Test for Geoengineering? Stratospheric Geoengineering Cannot be Tested in the Atmosphere without Full-Scale Implementation." *Science*, 327 (5965): 530–1.

Romanis, Elizabeth Chloe (2020). "Artificial Womb Technology and Clinical Translation: Innovative Treatment or Medical Research?" *Bioethics*, 34 (4): 392–402.

Romanis, Elizabeth Chloe (2023). "The Ethical and Legal Status of 'Fetonates' Or 'Gestatelings.'" *The American Journal of Bioethics*, 23 (5): 90–2.

Roose, Kevin (2023). " 'I think We're Heading Toward the Best World Ever': An Interview with Sam Altman." *The New York Times*, November 20, 2023. https://www.nytimes.com/2023/11/20/podcasts/hard-fork-sam-altman-transcript.html.

Rosen, Christine (2003). "Why Not Artificial Wombs? On the Meaning of Being Born, Not Incubated." *The New Atlantis*, 3 (Fall): 67–76.

Russell, Stuart (2022). "Artificial Intelligence and the Problem of Control." In Hannes Werthner, Erich Prem, Edward A. Lee, and Carlo Ghezzi (eds.), *Perspectives on Digital Humanism*, 19–24. Cham: Springer.

Salisbury, Mark (2023). "How Safe Are Self-Driving Cars as a Realistic Future for Our Roads?" *Fleet Point*, October 24, 2023. https://www.fleetpoint.org/autonomous-vehicles/how-safe-are-self-driving-cars-as-a-realistic-future-for-our-roads-2/.

Saltman, Max (2023). "As Azerbaijan Claims Final Victory in Nagorno Karabakh, Arms Trade with Israel Comes Under Scrutiny." *CNN*, October 4, 2023. https://www.cnn.com/2023/10/04/middleeast/azerbaijan-israel-weapons-mime-intl/index.html.

Sandel, Michael J. (2007). *The Case against Perfection: Ethics in the Age of Genetic Engineering*. Cambridge, MA: The Belknap Press of Harvard University Press.

Sandler, Ronald L. (2014). "Introduction: Technology and Ethics." In Ronald L. Sandler (ed.), *Ethics and Emerging Technologies*, 1–23. New York: Palgrave McMillan.

Sands, Leo (2022). "Drug Smuggling: Underwater Drones Seized by Spanish Police." *BBC*, July 4, 2022. https://www.bbc.com/news/world-europe-62040790.

Sarkin, Jeremy Julian, and Saba Sotoudehfar (2024). "Artificial Intelligence and Arms Races in the Middle East: The Evolution of Technology and its Implications for Regional and International Security." *Defense & Security Analysis*, 40 (1): 97–119.

Satariano, Adam, and Paul Mozur (2024). "AI Begins Ushering in an Age of Killer Robots." *The New York Times*, July 6, 2024. https://www.nytimes.com/2024/07/02/technology/ukraine-war-ai-weapons.html.

Shahzad, Qaiser (2007). "Playing God and the Ethics of Divine Names: An Islamic Paradigm for Biomedical Ethics." *Bioethics*, 21 (8): 413–8.

Shapiro, Zachary E. (2018). "Savior Siblings in the United States: Ethical Conundrums, Legal are Regulatory Void." *Washington and Lee Journal of Civil Rights and Social Justice*, 24 (2): 419–61.

Sharkey, Amanda (2019). "Autonomous Weapons Systems, Killer Robots and Human Dignity." *Ethics and Information Technology*, 21 (2): 75–87.

Shirouzu, Norihiko, and Chris Kirkham (2024). "Tesla's Robotaxi Push Hinges on 'Black Box' AI Gamble." *Reuters*, October 10, 2024. https://www.reuters.com/technology/tesla-gambles-black-box-ai-tech-robotaxis-2024-10-10/.

Shumailov, Ilia, Zakhar Shumaylov, Yiren Zhao, Yarin Gal, Nicolas Papernot, and Ross Anderson (2024). "The Curse of Recursion: Training on Generated Data Makes Models Forget." https://arxiv.org/abs/2305.17493.

Simons, Jake Wallis (2023). "Snapchat Is Playing God with Our Children's Wellbeing." *The Spectator*, May 1, 2023. https://www.spectator.co.uk/article/snapchat-is-playing-god-with-our-childrens-wellbeing/.

Simons, Massimiliano (2022). "Playing God: Symbolic Arguments against Technology." *Nanoethics*, 16 (2): 151–65.

Singal, Gaurav, and Anupam B. Jena (2024). "Opinion: It's Not Just Hype. AI Could Revolutionize Diagnosis in Medicine." *Los Angeles Times*, July 9, 2024. https://www.latimes.com/opinion/story/2024-07-09/artificial-intelligence-medicine-diagnosis.

Slattery, Peter, Alexander K. Saeri, Emily A. C. Grundy, Jess Graham, Michael Noetel, Risto Uuk, James Dao, Soroush Pour, Stephen Casper, and Neil Thompson (2024). "The AI Risk Repository: A Comprehensive Meta-Review, Database, and Taxonomy of Risks From Artificial Intelligence." *MIT AI Risk Repository*. https://airisk.mit.edu/ (accessed September 27, 2024).

Smith, Patrick Taylor (2018). "Legitimacy and Non-Dominion in Solar Radiation Management Research." *Ethics, Policy & Environment*, 21 (3): 341–61.

Solman, Paul (2024). "As Artificial Intelligence Rapidly Advances, Experts Debate Level of Threat to Humanity." *PBS New Hour*, August 23, 2024. https://www.pbs.org/newshour/show/as-artificial-intelligence-rapidly-advances-experts-debate-level-of-threat-to-humanity.

Solon, Olivia (2017). "Artificial Intelligence is Ripe for Abuse, Tech Researcher Warns: 'A Fascist's Dream'." *The Guardian*, March 13, 2017. https://www

.theguardian.com/technology/2017/mar/13/artificial-intelligence-ai-abuses-fascism-donald-trump.

Sovacool, Benjamin K., Chad Baum, and Sean Low (2023). "The Next Climate War? Statecraft, Security, and Weaponization in the Geopolitics of a Low-Carbon Future." *Energy Strategy Reviews*, 45: 101031.

Spence, James H (2008). "What Is Wrong with Cloning a Dinosaur? Jurassic Park and Nature as a Source of Moral Authority." In Dean A. Kowalski (ed.), *Steven Spielberg and Philosophy: We're Gonna Need a Bigger Book*, 97–111. Lexington: The University of Kentucky Press.

Stein, Rob (2024). "An Artificial Womb Could Build a Bridge to Health for Premature Babies." *NPR*, April 12, 2024. https://www.npr.org/sections/health-shots/2024/04/12/1241895501/artificial-womb-premature-birth.

Straughan, David (2023). "Are Self-Driving Cars a National Security Risk?" *The Seattle Medium*, July 27, 2023. https://seattlemedium.com/are-self-driving-cars-a-national-security-risk/.

Sun, Yalin, and Yan Zhang (2021). "A Review of Theories and Models Applied in Studies of Social Media Addiction and Implications for Future Research." *Addictive Behaviors*, 114: 106699.

Supramaniam, P. R., and M. Mittal (2018). "Assisted Reproductive Technology: Where Did the Journey Begin, and Where are We Today—A 40-Year History." *World Journal of Gynecology & Women's Health*, 1 (4): 1–3.

Taddeo, Mariarosaria, and Alexander Blanchard (2022). "Accepting Moral Responsibility for the Actions of Autonomous Weapons Systems—A Moral Gambit." *Philosophy & Technology*, 35 (3): 1–24.

Taeihagh, Araz, and Hazel Si Min Lim (2019). "Governing Autonomous Vehicles: Emerging Reponses for Safety, Liability, Privacy, Cybersecurity, and Industry Risks." *Transport Reviews*, 39 (1): 103–28.

Tenbarge, Kat (2024). "The Defiance Act Passes in the Senate, Potentially Allowing Deepfake Victims to Sue over Nonconsensual Images." *NBC News*, July 24, 2024. https://www.nbcnews.com/tech/tech-news/defiance-act-passes-senate-allow-deepfake-victims-sue-rcna163464.

Ulgen, Ozlem (2016). "Human Dignity in an Age of Autonomous Weapons: Are We in Danger of Losing an 'Elementary Consideration of Humanity'?" *European Society of International Law Conference Paper Series*, 8 (9): 1–19.

Umbrello, Steven, Phil Torress, and Angelo F. De Bellis (2020). "The Future of War: Could Lethal Autonomous Weapons Make Conflict More Ethical?" *AI & Society*, 35 (1): 273–82.

UN News (2019). "Autonomous Weapons That Kill Must be Banned, Insists UN Chief." *UN News: Culture and Education*, March 25, 2019. https://news.un.org/en/story/2019/03/1035381.

United Nations Environment Programme (2023). "One Atmosphere: An Independent Expert Review on Solar Radiation Modification Research and Deployment." *United Nations Environment Programme*, February 28, 2023. https://www.unep.org/resources/report/Solar-Radiation-Modification-research-deployment.

U.S. Department of Health and Human Services (2024). "Fact Sheet: In Vitro Fertilization (IVF) Use Across the United States." *U.S. Department of Health and Human Services*, March 13, 2024. https://www.hhs.gov/about/news/2024/03/13/fact-sheet-in-vitro-fertilization-ivf-use-across-united-states.html.

The U.S. Surgeon General's Advisory (2023). "Social Media and Youth Mental Health." *U.S. Department of Health and Human Services*, May 23, 2023. https://www.hhs.gov/sites/default/files/sg-youth-mental-health-social-media-advisory.pdf.

Tulp, Sophia (2022). "Video of 'Artificial Womb' Concept Doesn't Show Real Facility." *The Associated Press*, December 23, 2022. https://apnews.com/article/fact-check-artificial-womb-simulation-video-348593206360.

Van den Belt, Henk (2009). "Playing God in Frankenstein's Footsteps: Synthetic Biology and the Meaning of Life." *NanoEthics*, 3 (3): 257–68.

Varga, David (2023). "Ilya: The AI Scientist Shaping the World." *Effective Altruism Forum*, November 20, 2023. https://forum.effectivealtruism.org/posts/THfXNTP6YdXnNge8P/ilya-the-ai-scientist-shaping-the-world.

Vassilev, Apostol, Alina Oprea, Alie Fordyce, and Hyrum Andersen (2024). "Adversarial Machine Learning: A Taxonomy and Terminology of Attacks and Mitigations." *NIST Trustworthy and Responsible AI, National Institute of Standards and Technology*. Gaithersburg, MD. https://doi.org/10.6028/NIST.AI.100-2e2023.

Vaughn, Lewis (2023). *Bioethics: Principles, Issues, and Cases*, 5th ed. New York: Oxford University Press.

Veatch, Robert M. (2016). *The Basics of Bioethics*, 3rd ed. New York: Routledge.

Verma, Pranshu, and Cat Zakrzewski (2024). "AI Deepfakes Threaten to Upend Global Elections. No One Can Stop Them." *The Washington Post*, April 23, 2024. https://www.washingtonpost.com/technology/2024/04/23/ai-deepfake-election-2024-us-india/.

Vosoughi, Soroush, Deb Roy, and Sinan Aral (2018). "The Spread of True and False News Online." *Science*, 359: 1146–51.

Walton, Douglas, Christopher Reed, and Fabrizio Macagno (2008). *Argumentation Schemes*. New York: Cambridge University Press.

Waytz, Adam and Liane Young (2019). "Aversion to Playing God and Moral Condemnation of Technology and Science." *Philosophical Transactions of the Royal Society, B* 374 (1771): 20180041.

Weckert, John (2016). "Playing God: What is the Problem?" In Steve Clarke, Julian Savulescu, C. A. J. Coady, Alberto Giubilini, and Sagar Sanyal (eds.), *The Ethics of Human Enhancement: Understanding the Debate*, 87–99. New York: Oxford University Press.

Weigle, Paul E., and Reem M. A. Shafi (2024). "Social Media and Youth Mental Health." *Current Psychiatry Reports*, 26: 1–8.

Weitzman, Martin L. (2015). "A Voting Architecture for the Governance of Free-Driver Externalities with Application to Geoengineering." *The Scandinavian Journal of Economics*, 117 (4): 1049–68.

Wierenga, Edward R. (1989). *The Nature of God: An Inquiry into Divine Attributes*. Ithaca: Cornell University Press.

Wilkinson, Stephen, Nicola J. Williams, and Sara Fovargue (2023). "Artificial Wombs Could Someday Be a Reality—Here's How They May Change Our Notions of Parenthood." *The Conversation*, December 1, 2023. https://theconversation.com/artificial-wombs-could-someday-be-a-reality-heres-how-they-may-change-our-notions-of-parenthood-217490.

Wong, Pak-Hang (2015). "Confucian Environmental Ethics, Climate Engineering, and the 'Playing God' Argument." *Zygon: Journal of Religion & Science*, 50 (1): 28–41.

Wong, Queenie, and Wendy Lee (2024). "Trump Posted a Fake Taylor Swift Image. AI and Deepfakes Are Only Going to Get Worse this Election Cycle." *Los Angeles Times*, August 21, 2024. https://www.latimes.com/entertainment-arts/business/story/2024-08-21/deep-fakes-social-media-taylor-swift-donald-trump.

Wood, Nathan Gabriel (2020). "The Problem with Killer Robots." *Journal of Military Ethics*, 19 (3): 220–40.

Wood, Nathan Gabriel (2023). "Autonomous Weapon Systems and Responsibility Gaps: A Taxonomy." *Ethics and Information Technology*, 25: 16.

Yoshida, Junko (2015). "Self-Driving Cars: 'My Program Made Me Do It.'" *EE Times*, October 23, 2015. https://www.eetimes.com/self-driving-cars-my-program-made-me-do-it/.

Zarnetskea, Phoebe, L., Jessica Gurevitch, Janet Franklind, Peter M. Groffmane,Cheryl S. Harrisong, Jessica J. Hellmannh, Forrest M. Hoffman, Shan Kotharil, Alan Robockn, Simone Tilmeso, Daniele Visionip, Jin Wuq, Lili Xia, and Cheng-En Yang (2021). "Potential Ecological Impacts of Climate Intervention by Reflecting Sunlight to Cool Earth." *Proceedings of the National Academy of Sciences of the United States of America*, 118 (15): e1921854118.

Zialcita, Paolo (2019). "Facebook Pays $643,000 Fine for Role in Cambridge Analytica Scandal." *NPR*, October 30, 2019. https://www.npr.org/2019/10/30/774749376/facebook-pays-643-000-fine-for-role-in-cambridge-analytica-scandal.

Zimmer, Marc (2005). *Glowing Genes: A Revolution in Biotechnology*. New York: Prometheus Books.

Index

About the Author

Moti Mizrahi is Professor of Philosophy, with expertise in the philosophy of science and technology, at the Florida Institute of Technology, where he teaches Artificial Intelligence (AI) Ethics, Bioethics, Engineering Ethics, and Philosophy of Science regularly. He received his PhD in philosophy from the Graduate Center of the City University of New York in 2010. His books include *The Kuhnian Image of Science: Time for a Decisive Transformation?* (Rowman & Littlefield, 2018), *The Relativity of Theory: Key Positions and Arguments in the Contemporary Scientific Realism/Antirealism Debate* (Springer, 2020), and *For and Against Scientism: Science, Methodology, and the Future of Philosophy* (Rowman & Littlefield, 2022).

www.ingramcontent.com/pod-product-compliance
Ingram Content Group UK Ltd.
Pitfield, Milton Keynes, MK11 3LW, UK
UKHW020209290926
15470UKWH00001B/83

9 781666 964523